Stars as Fossils: Galactic History

Nama

Contents

Chapter 1

Introduction

This chapter introduces the physical and observational properties of white dwarfs and how they can be used as tracers for Galactic formation and evolution. This chapter begins with a bit of history on the discovery and understanding of white dwarfs in Section 1.1. Section 1.2 describes the physical and observational characteristics of white dwarfs, including their presence in photometric surveys and the information that can be acquired as part of follow-up spectroscopy. Section 1.3 details how white dwarfs can be used to study their host stellar population, including the luminosity function, age, and star formation history. Section 1.4 focuses on these properties related to white dwarfs in the Galactic halo, which remain elusive in modern surveys due to their rarity and intrinsic faintness. The chapter finishes with an overview of the remainder of this book, including the motivations for the selected projects.

1.1 White Dwarfs and Stellar Evolution

The population of stars that form white dwarfs account for roughly 97% of all stars in the Milky Way and hence their properties provide valuable insight into its formation and evolution. The term 'white dwarf' was first used in 1922 by astronomer Willem Luyten to describe the peculiar stars found orbiting other popular nearby stars. Much of the early focus was on Sirius B (seen in Figure 1.1) as it is the closest white dwarf to the Sun, a mere 8.6 light-years away.

Sirius B was discovered in 1862 by Alvin Clark following its proposal by Friedrich Bessel in 1844 based on variations in the proper motion of Sirius A (Bessel, 1844; Holberg & Wesemael, 2007). As Figure 1.1 shows, Sirius B is much fainter than its A-type main-sequence companion, Sirius A. Despite this luminosity difference, both stars were noted for their similar blue colour. In fact, initial observations of Sirius

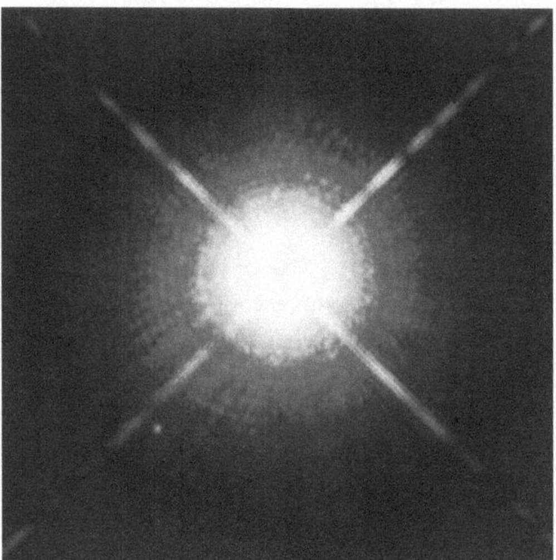

Figure 1.1 NASA image of Sirius A and its faint companion, Sirius B (lower-left). The separation between the two stars is roughly 5 arcseconds. This image was taken using the Wide Field and Planetary Camera 2 (WFPC2) aboard the Hubble Space Telescope. Image credit: NASA, ESA, H. Bond (STScI), and M. Barstow (University of Leicester).

B led many to assume that it merely reflected the light from its companion and was therefore not self-luminous (Holberg, 2009).

The mass ratio between Sirius A and Sirius B was determined by Otto Struve in 1866, who found that Sirius B was roughly half the mass of Sirius A, despite the fact that Sirius B is more than 10,000 times fainter. This led to the conclusion that the two stars "are of a very different constitution" (Struve, 1866).

Observations in the early 1900s determined that Sirius B, as well as another nearby white dwarf, 40 Eri b, had higher temperatures than many main-sequence stars, even Sirius A. A spectrum was finally acquired in 1915 by Walter Adams, revealing that it was nearly identical to its A-type companion (Adams, 1915). Combining all of these observations meant that Sirius B had a mass comparable to that of the Sun and a radius comparable to the Earth, resulting in a density one million times higher than our Sun. This corroborated Struve's assertion that these stars were formed in very different manners, despite their similar spectra.

The theoretical basis for the unusual interiors of white dwarfs was proposed in 1926 by Ralph Fowler, who postulated that the interior could be supported by electron degeneracy pressure (Fowler, 1926; Holberg, 2009). This theory was expanded upon by Subrahmanyan Chandrasekhar, who showed that white dwarfs have a maximum possible mass as a consequence of the degenerate core before the gravitational force would overpower the force due to electron degeneracy pressure Chandrasekhar (1931).

The work by these pioneering astronomers set the stage for early white dwarf research. The fact that there existed a handful of white dwarfs close to the Sun suggested that these objects were quite numerous throughout the Milky Way. The difficulty, however, lies in their intrinsic faintness that required the advent of large wide-field surveys to discover large samples. These samples have allowed white dwarfs to be used to study stellar populations as a whole. The following sections will describe more recent advances made using white dwarfs as they relate to stellar and Galactic evolution.

1.1.1 Formation

White dwarfs represent the end stage of stellar evolution for all low- and intermediate-mass stars ($M \lesssim 8\ M_\odot$). For a star like our Sun, its lifetime is dominated by the process of fusing hydrogen into helium within its core. This evolutionary stage is called the main sequence, regardless of mass, and this phase can last anywhere from a few Myr in the most massive stars to many tens of Gyrs in the least massive stars (Hurley et al., 2000).

The post-main-sequence evolution is largely determined by a star's initial mass. Figure 1.2 presents a schematic overview for a star with an initial mass equal to our Sun. Following the exhaustion of hydrogen the core will begin to contract as the pressure generated by the core-burning will cease. The core will continue to contract as a shell of hydrogen burning commences surrounding the core. This causes the atmosphere to expand and as such the star enters the Red Giant phase. The star begins increasing in luminosity as the outer layers expand and cool. As the inert helium core collapses, its temperature will increase until it becomes hot enough to ignite via the triple-alpha process. The triple-alpha process is the means for which one carbon atom is formed by fusing three helium atoms together. The star will then enter a stable period where it burns helium into carbon and oxygen in the core, called the horizontal branch. Following the exhaustion of helium in the core, it begins to collapse in a similar way to its evolution up the red giant branch. This phase is called the asymptotic giant branch (AGB) phase, and it is characterized by concentric shells of helium and hydrogen burning surrounding an inert carbon-oxygen core. Stars like the Sun do not have enough mass to contract to the point of carbon ignition, and therefore the AGB phase ends when the hydrogen and helium shells are depleted. The star begins to shed the outer envelope, resulting in the formation of a planetary nebula that surrounds the remnant carbon-oxygen core — a white dwarf.

1.1.2 Evolution

The interior of a white dwarf is supported by electron degeneracy pressure since it lacks the temperatures needed to generate radiation pressure through fusion. Electron degeneracy pressure is a result of the Pauli Exclusion principle that prohibits two fermions from occupying identical quantum states. As matter begins to collapse, the fermions will begin to exert a pressure as the lowest quantum states begin to fill up.

Since white dwarfs do not undergo nuclear fusion in their core, their evolution is dominated by the slow loss of thermal energy, dubbed a "cooling curve" (see Figure 1.3). The energy loss comes initially from the release of neutrinos through the decay of heavier elements left over from the AGB phase. Later, energy losses are dominated by thermal processes and the crystallization of the carbon interior (Lamb & van Horn, 1975). Due to the small surface areas of these objects the cooling is very slow and

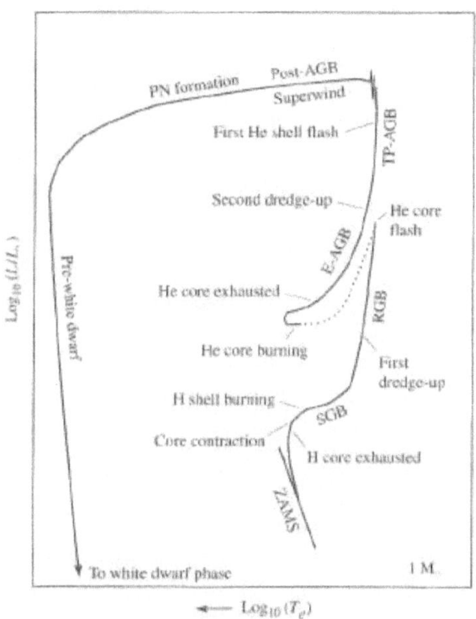

Figure 1.2 The evolution of a $1\,\mathrm{M}_\odot$ star from the zero-age main-sequence (ZAMS) to the beginning of the white dwarf phase. Figure taken from Carroll & Ostlie (2006).

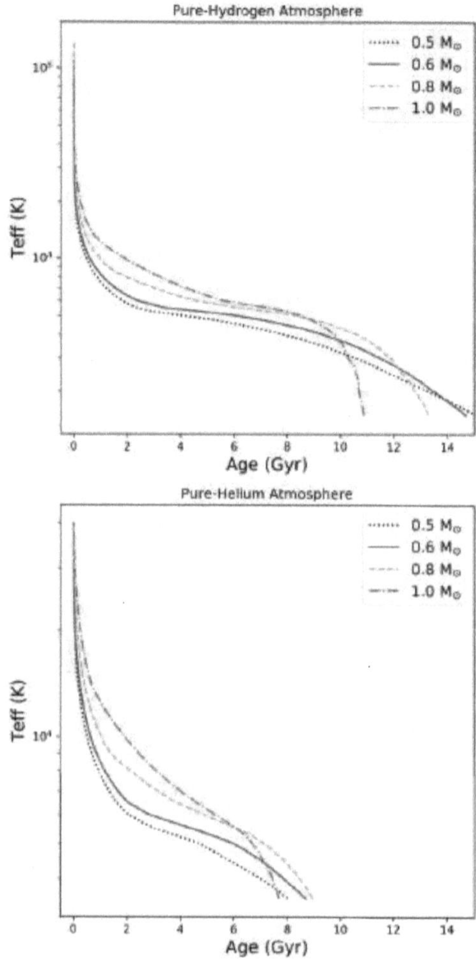

Figure 1.3 A cooling curve, for pure H-atmosphere (top) and pure He-atmosphere (bottom) white dwarfs. The model curves are presented in Holberg & Bergeron (2006) and show the decrease in temperature as a function of time in Gyr.

can last much longer than the current age of the universe. Thus, the age of a stellar population can be determined using the coolest white dwarfs (see, e.g, Bergeron et al., 1997; Harris et al., 2006; Kilic et al., 2017).

1.2 White Dwarf Properties

1.2.1 Photometry

White dwarfs have historically been difficult to detect due to their faint luminosities. This meant that deep, wide field, surveys are required to observe a large sample of objects. The arrival of surveys, such as the Luyten Half-Second Survey (Luyten, 1979) and more recently the Sloan Digital Sky Survey (SDSS; York et al., 2000), the Canada France Imaging Survey (CFIS; Ibata et al., 2017), and *Gaia* (Gaia Collaboration et al., 2018b), have uncovered on the order of 10^5 white dwarf candidates. This is typically done using a combination of colours and proper motion measurements, as the unique combination of low luminosities and high surface temperatures make young, hot, white dwarfs relatively easy to select. Typically, the use of a blue filter, such as a u-band, is employed as it is sensitive to the Balmer lines present in the majority of white dwarfs. This can be seen in the top panel of Figure 1.4, which shows the location of various white dwarf types along with model cooling curves for a 0.6 $M_\odot$ pure H-atmosphere and He-atmosphere in a colour-colour diagram.

Selecting white dwarfs on photometry alone becomes difficult as they cool into the same temperature region as main-sequence stars. This regime also represents the temperature where the intensity of the hydrogen lines decreases. In order to select white dwarfs of all temperatures one of their main characteristics can be exploited: their faintness means that they are relatively close to the Sun, and thus experience a larger proper motion compared to other stars at equal magnitude. By combining the photometry and proper motion, the reduced proper motion diagram (RPMD) is created and is used to cleanly select white dwarfs of all temperatures (see, e.g, Harris et al., 2006; Rowell & Hambly, 2011; Fantin et al., 2017; Munn et al., 2017). This separation can easily be seen in the bottom panel of Figure 1.4, where spectroscopically classified white dwarfs are highlighted in blue.

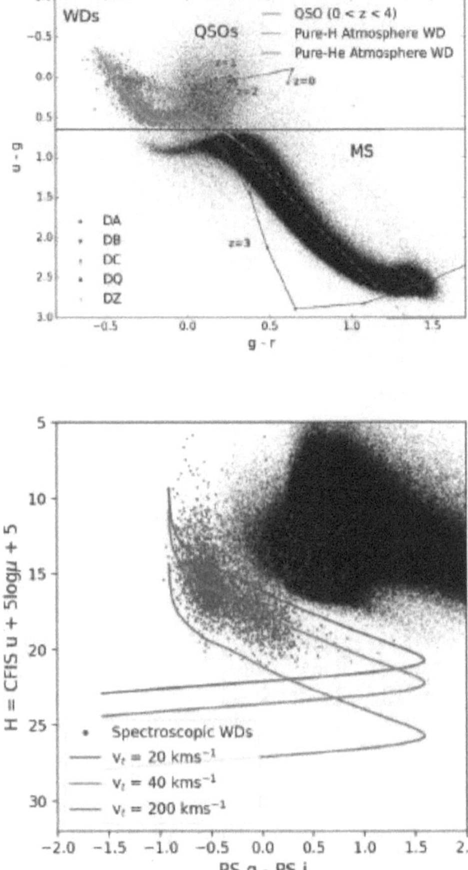

Figure 1.4 *Top*: Colour-colour diagram showing the location of various white dwarf types. Also plotted are model cooling curves for a 0.6 M$_\odot$ pure H-atmosphere (green) and pure He-atmosphere (blue). Figure taken from Ibata et al. (2017). *Bottom*: Reduced Proper Motion Diagram for CFIS objects, with spectroscopic white dwarfs from Kleinman et al. (2013) highlighted in blue, showing the separation between the white dwarfs and main-sequence stars over all temperatures.

1.2.2 Spectroscopy

Since white dwarfs represent the final evolutionary stage for the vast majority of stars their masses are an important property for which theories of stellar evolution can be tested against (Koester et al., 1979). Specifically, the mass of a white dwarf can provide information on the mass loss during the post-main-sequence evolution of a star, as well as any effects of binary interactions (Liebert et al., 2005). The upper mass limit, a consequence of electron degeneracy pressure first noted by Chandrasekhar (1931), has also been instrumental in our study of type 1a supernovae and their use as standard candles (Perlmutter et al., 1999). Furthermore, the mass and temperature are needed to derive an age, which is important for studying the evolution of their stellar population.

The evolution of a white dwarf is dependent on its mass and atmospheric composition, as shown in Figure 1.3, both of which can be determined with spectroscopy. Typically white dwarf spectra are composed of just a single element (Liebert, 1980), with greater than 80% of local white dwarfs containing a hydrogen atmosphere (Limoges et al., 2015). White dwarf spectra are classified based on the dominant spectral features. Those that show only strong hydrogen lines are classified as type DA and those with only strong He I lines as type DB. There are, however, many different types, a few of which can be seen in Figure 1.5. These include white dwarfs with only strong carbon lines (DQ), other metal-lines (DZ), He II lines (DO), and those without any spectral features (DC). White dwarfs can also have mixed atmospheres and the classification will contain the letters associated with the dominant spectra feature followed by any other less dominant features. An example is the type DBA, which is a white dwarf with an atmosphere dominated by helium with only traces of hydrogen. Statistically, hydrogen-rich or helium-rich atmospheres account for more than 99% of all white dwarfs, and therefore are typically the focus of large scale studies (Kepler et al., 2007, 2016a).

The surface gravity and temperature can be determined by fitting the observed spectrum with model atmospheres (see, e.g, Bergeron et al., 1994; Kleinman et al., 2013; Kepler et al., 2015; Dame et al., 2016). This is because the width and depth of the spectral lines are a function of surface gravity and temperature. Given the high surface gravities of white dwarfs, the lines are much broader than main-sequence stars.

The surface gravity, $\log g$, can be turned into a mass using the mass-radius relation

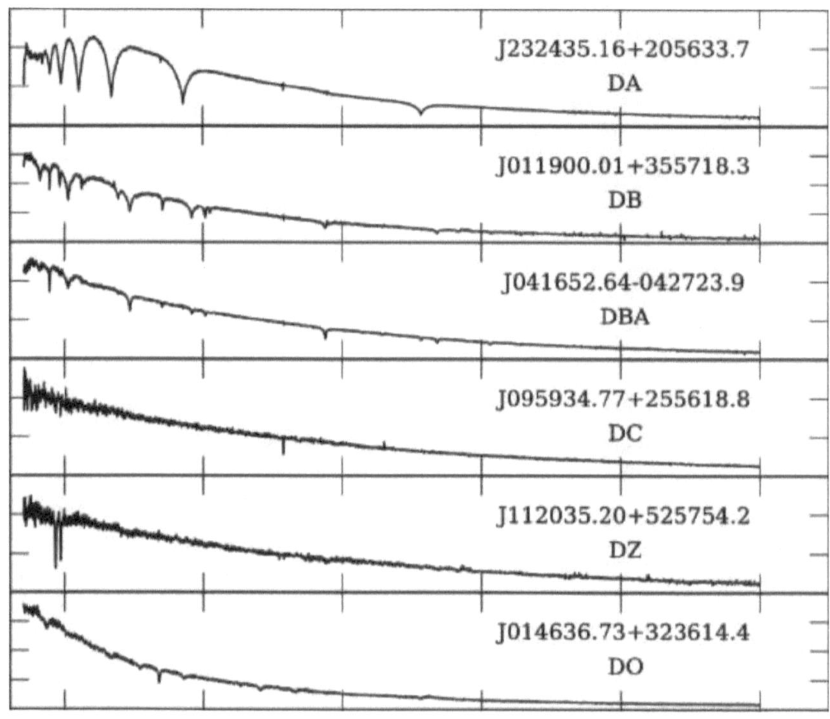

Figure 1.5 A sample of white dwarfs with differing atmospheric compositions: hydrogen (DA), helium (DB), hydrogen+helium (DBA), featureless (DC), metal lines (DZ), and He II (DO). Figure taken from Gentile Fusillo et al. (2015).

(MRR). The first derivation of this relationship was done by Chandrasekhar (1931), and states that R $\sim$ M$^{-1/3}$, a direct consequence of being a degenerate gas. Minor corrections have been made to account for electrostatic forces (at low densities), and beta decay (at high densities), which alter the mean molecular weight of the gas (Hamada & Salpeter, 1961). The MRR has been extensively tested by comparing masses obtained with parallaxes to those obtained spectroscopically (see, e.g, Schmidt, 1996; Vauclair et al., 1997; Tremblay et al., 2017; Joyce et al., 2018), and found to be consistent with observations.

The mass of a white dwarf is particularly important when determining the age, as both the cooling age and progenitor age are strong functions of mass. For example, at 10,000 K the cooling age can vary by about 3 Gyr depending on the mass, and by 3,000 K the difference is on the order of the age of the Universe. Despite the inverse relationship between the mass and radius, higher mass white dwarfs cool faster. This is because more massive white dwarfs have lower heat capacities due to their increased densities. This results in strong vibrations in the crystal lattice in the core which promotes heat loss (Renedo et al., 2010; Isern & García–Berro, 2004). This process is called Debye cooling.

By applying the MRR to spectroscopic observations from the SDSS, Kepler et al. (2007) showed that the majority ($\sim$75%) of white dwarfs have a mass of roughly 0.6 M$_\odot$. An excess of white dwarfs with lower and higher masses are thought to form through binary interactions (Calcaferro et al., 2018) and mergers (Cheng et al., 2019) respectively. This distribution can be seen in the left-hand panel of Figure 1.6. The temperature distribution for DAs can be seen in the right-hand panel and shows the increased number of cool white dwarfs resulting from the slowing of the cooling process with time (D'Antona & Mazzitelli, 1990).

1.3   White Dwarfs and the Milky Way Disk

Most of the stars in the Milky Way are located in a rotating disk centered in the Galactic plane. The disk can be decomposed into two components. The first is a young thin disk, which has a smaller scale height ($\sim$300 pc), a higher metallicity, and space velocities similar to our Sun. The thick disk, which has a larger scale height than the thin disk ($\sim$800 pc), contains older stars with lower iron-abundances, enhanced alpha-abundances, and larger space velocities relative to the thin disk (Bensby et al., 2003; Haywood et al., 2013; Robin et al., 2017). This section details studies which

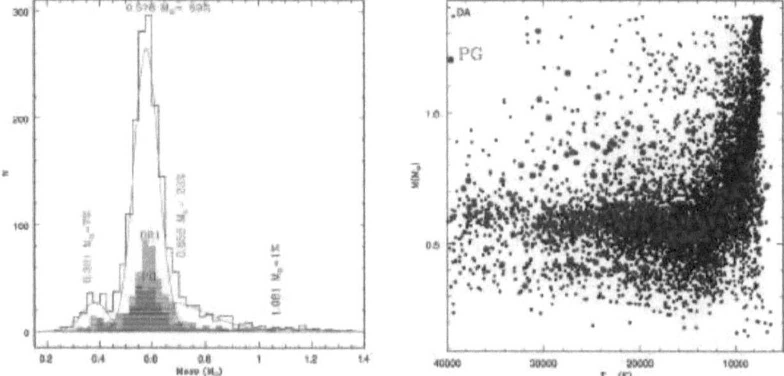

Figure 1.6 *Left:* The mass distribution of white dwarfs in SDSS DR7. The distribution is peaked at roughly 0.6 $M_\odot$, with smaller contributions from low- and high-mass components. *Right:* The temperature distribution for the same sample, showing the build-up of white dwarfs at low temperatures where the cooling process slows down. Figure taken from Kepler et al. (2007).

have used white dwarfs to measure the ages of these two components.

1.3.1   Age and the White Dwarf Luminosity Function

The age of a star, or a stellar population in general, is a fundamental astrophysical variable as it allows astronomers to understand both stellar and galactic evolution. Unfortunately, the age of a star is not directly measurable and therefore requires a model. This will inherently introduce systematic uncertainties, and many main-sequence models are not accurate over a broad range of ages and metallicities (Soderblom, 2010).

Due to their ubiquitous nature and well-established evolution white dwarfs can provide a measurement of the age of a stellar population. Historically, this has been done using the white dwarf luminosity function (WDLF) – the number density of white dwarfs as a function of bolometric magnitude. If, for example, a population was formed from a single burst of star formation, you would expect the luminosity function to increase monotonically until it drops off abruptly as the cooling ages of the white dwarfs approach the age of the population (see Figure 1.7). Thus, by studying the coolest white dwarfs in a given stellar population one can determine its age.

There are, however, a few systematic uncertainties associated with determining the age of a white dwarf. The largest source of uncertainty is the initial-to-final mass relation (IFMR) which relates the progenitor mass to the resulting white dwarf mass. This relation has been calibrated using star clusters as they typically contain a single stellar population for which initial and white dwarf masses can be determined. The relation has been found to be nearly linear with minimal scatter, however, the relation has not been well calibrated for low mass progenitors. This is because the objects needed for this study — old globular clusters — have very faint white dwarfs (Kalirai et al., 2008). The uncertainties introduced by the IFMR can be minimized if one only studies the oldest white dwarfs, as their progenitor age accounts for a smaller fraction of the total age of the star.

This method has been applied to many globular and open clusters, which have typically been associated with a short burst of star formation. This method provides an independent age measurement since the age of these clusters is otherwise determined by comparing theoretical isochrones consisting of the main-sequence and red-giant branch to the observed colour-magnitude diagram. Using the white dwarf method, ages of globular clusters have been found to be ~10-13 Gyr depending on

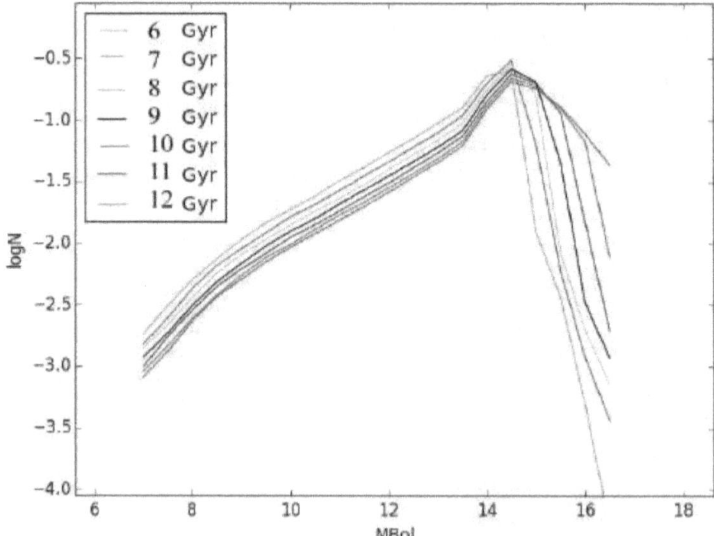

Figure 1.7 Model luminosity functions for single burst stellar populations of various ages. As the age increases, the relative number of white dwarfs beyond the turn-off increases as these objects have had more time to cool.

metallicity (Hansen et al., 2004, 2007, 2013). Open clusters that are typically associated with the disk have been found to have ages of $\sim$ 2-9 Gyr Bedin et al. (2008, 2015).

Cluster ages are much easier to determine as they contain, roughly, a single burst of star formation and the stars all lie at nearly the same distance. The Milky Way's disk, on the other hand, continues to form stars at a variety of distances. This makes luminosity measurements for white dwarfs difficult. The first such study of the Milky Way's disk was done by Winget et al. (1987), who modelled the WDLF from the Luyten Half-Second proper motion catalogue to obtain an age measurement of 9.3 ± 2.0 Gyr for the Galactic disk and 10.3 ± 2.2 Gyr for the age of the Universe. These early studies, however, suffered from strong incompleteness and a lack of faint objects due to their shallow photometric data.

Recent studies by Harris et al. (2006), Rowell & Hambly (2011), and Munn et al. (2017) have taken advantage of deeper wide-field surveys to construct detailed luminosity functions for the Galactic components. These surveys have revealed a nearly monotonically increasing WDLF, consistent with a nearly constant star formation history, followed by a steep drop at roughly $M_{bol} \sim 15.3$. Rowell & Hambly (2011) were able to decompose their luminosity function into a thin and thick disk component by introducing a statistical weight based on the tangential velocity of each star implied from the RPMD. Kilic et al. (2017) took this study a step further by simultaneously fitting the observed WDLF from Munn et al. (2017) with a model thin and thick disk component. The resulting ages for the disk components were found to be 7.8 ± 0.4 Gyr and 9.7 ± 0.2 Gyr for the thin and thick disk, respectively.

1.3.2 Star Formation History

The star formation history (SFH) of a given galaxy provides important information regarding the build-up of mass, including both its past merger history and secular processes. High redshift studies have revealed that more than 50% of the stellar mass contained within a Milky Way-like galaxy was formed more than 8 Gyr ago ($z \gtrsim 1$) (van Dokkum et al., 2013), and that the star formation rate (SFR) drops over time as the gas reservoir is depleted. Since many of the bright stars born very early in the Universe no longer exist, the study of the SFH in the Milky Way is typically done by studying the signatures left by these early stars. One such method is to study the Galactic chemical evolution (GCE) which uses either an open- (allowing for gas accretion) or closed-box (no gas accretion) and convert the gas density into a SFR. The resulting metallicity distributions are then compared to observations.

A recent comprehensive study of the SFH in the inner Milky Way was performed by Haywood et al. (2018) using the GCE model of Snaith et al. (2015). The resulting SFH implies a rapid build-up of mass within the first $3 - 4$ Gyr followed by a period of negligible star formation before returning to a roughly constant rate over the past $7.5 - 8$ Gyr. These two periods are typically associated with a thin and thick disk and imply that the two components are roughly equal in total mass.

This result is at odds with many recent and classical results, which limit the contribution by the thick disk to the total stellar mass. Many classical studies assumed that the thick disk could contribute no more than 25% of the total mass, with measurements varying from $5 - 25\%$ (see, e.g. Chang et al., 2011; McMillan, 2011; Micali et al., 2013). On the other hand, results from other local disk galaxies show examples that include a thin and thick disk with roughly equal mass. For example, Comerón et al. (2011) used Spitzer data of nearby edge-on spiral galaxies and concluded that this scenario is quite typical in the local Universe.

The shape of the WDLF also contains information regarding the star formation history of the Milky Way. Harris et al. (2006) concluded that their linearly increasing WDLF was the result of a roughly constant star formation rate (SFR) over the lifetime of the disk, except for a small increase in the past roughly 500 Myr. The SFH derived using white dwarfs has focused on the solar neighbourhood as correcting for completeness is difficult. Rowell (2013) performed the first such inversion of the WDLF, showing a bimodal SFR with a dip around 6 Gyr. These results were corroborated by Tremblay et al. (2014) who used a complete, volume-limited sample within 20 pc. The advantage of the local sample is that they have distances derived from parallax which can be converted into accurate masses, leading to more accurate total ages via the IFMR. The primary drawback to using the local white dwarfs is the small sample size. Furthermore, the small volume probed may not be representative of the disk. Thus, in order to expand this study a model that incorporates selection, survey, and Milky Way density effects would need to be accounted for.

1.4   Halo White Dwarfs

Halo white dwarfs have been a long sought after population as they can provide a direct measurement for the age of the Milky Way and the duration of halo star formation. However, the study of these objects has been hindered by their intrinsically faint luminosities and very low number densities in the solar neighbourhood. These objects are typically selected based on their large tangential velocities ($v_t > 200$ kms^{-1}) inferred through the RPMD (see the green line in the bottom panel of Figure 1.4). The first halo white dwarf candidates were presented by Liebert et al. (1989), who identified only six candidates from the all-sky Luyten Half-Second catalogue.

A study by Oppenheimer et al. (2001), who claimed that cool white dwarfs could account for at least 2% of the 'unseen' matter in the halo, sparked fierce debate over the nature of halo white dwarfs. In this study, the authors used the SuperCOSMOS survey to identify 38 halo candidates using an RPMD. The debate focused on the age of these halo white dwarfs, as many of them had cooling ages that were consistent with a disk origin if one assumed a white dwarf mass of 0.6 M$_\odot$. Bergeron (2003) and Bergeron et al. (2005) provided a detailed photometric study of these high-velocity white dwarfs and cited the need for spectroscopic measurements that provide accurate mass determinations in order to properly derive total ages – that is, the sum of the cooling and progenitor age.

The largest sample of spectroscopically confirmed halo white dwarfs for which accurate masses could be determined came from Kalirai (2012), who used four objects to obtain a mean age of 11.4 ± 0.7 Gyr for the inner halo. This value is consistent with Kilic et al. (2010) who found total ages of $10-11$ Gyr for three halo white dwarfs in the SDSS. A larger sample would provide a measurement of the duration of star formation history and a more accurate portrayal of the mean age of the inner halo. This was the motivation for Chapter 4, where I acquired a larger sample of halo white dwarfs. These candidates were selected using CFIS and Gaia with the goal of making the first measurement of the duration of star formation in the inner halo using white dwarfs.

Recent studies have uncovered larger samples of halo white dwarfs which can be followed up spectroscopically. The largest such photometric sample was compiled by Munn et al. (2017), who used second epoch observations of the $\sim$2000 deg^2 within the SDSS footprint to uncover 137 objects with reduced proper motions consistent with the halo. The luminosity function lacks an observed turnoff, and hence the age of the population remains to be determined. .

1.5 book Structure

As outlined above, historically white dwarfs have been used to study the formation and evolution of the Milky Way through the luminosity function. The main goal of this book is to present a complementary method by developing a white dwarf population synthesis model, which we motivate and describe in Chapter 2.

Such a model would allow for a study of the Milky Way's evolution, as changing the various inputs in the model would directly affect the resulting white dwarf population. In Chapter 3 we apply the model to the white dwarf sample in CFIS to derive the star formation history of the Milky Way.

One conclusion of this project was that the halo sample was poorly constrained due to the small sample size in the CFIS data. Consequently, I acquired spectroscopic follow-up for a number of CFIS halo white dwarf candidates to directly measure individual ages to determine the duration of star formation in the Galactic halo. This work is presented in Chapter 4 and describes an observational program where spectroscopic observations of 18 halo white dwarf candidates were obtained to measure their temperature and masses, both of which are needed to derive an age.

Since the sample of halo white dwarfs remains on the order of 150 (Munn et al., 2017; Kilic et al., 2019) we would like to know what potential sample sizes will be observed in upcoming large, wide-field surveys including Legacy Survey of Space and Time, the Roman Space Telescope's high latitude survey, Euclid, and CASTOR. These results are presented in Chapter 5 and show that future surveys will increase the sample size by many orders of magnitudes. This will open up a wide variety of future science cases, and we explore potential results of a few of these cases, including the IFMR, the halo luminosity function, and pulsating white dwarfs.

Finally, this book concludes with a holistic reflection of the work presented in the previous chapters. It then finishes by discussing the critical questions that remain to be answered in the field and how future studies will greatly aid in the quest to use white dwarfs as tracers for Galactic evolution.

Chapter 2

Modeling the Milky Way's White Dwarf Population

Much of this chapter was published as part of Fantin et al. (2019), Astrophysical Journal, Volume 887, Issue 1, pp. 148, with additional figures and descriptions added for context.

This chapter describes the motivation and development of a white dwarf population synthesis model. Section 2.1 describes previous population synthesis models that include white dwarfs, including their advantages and shortcomings. Section 2.2 details the model itself, including many assumptions about the Milky Way and white dwarf physics. I conclude in Section 2.3, which describes the outputs of the model and its potential applications.

2.1 Milky Way Stellar Population Synthesis Models and White Dwarfs

The ability to simulate star counts in a given survey is useful for testing theories of stellar and Galactic evolution since inputs to the model will produce differing outputs. These stellar population synthesis codes mainly focus on main-sequence and post-main-sequence stars and take various input physics, such as the lifetime, age distribution, metallicity, density distributions, and mass to produce a catalogue of stars with photometry from a defined survey.

While there exist many such models, the most popular versions include the Besançon (Robin et al., 2003, 2017) and TRILEGAL (Girardi et al., 2005) models. Besançon includes both DA and DB white dwarfs, however, they are manually added to the survey given an observed density from the SDSS. In other words, the objects are not allowed to form naturally through stellar evolution. As a result, the predicted counts do not provide any insights into Galactic evolution. TRILEGAL, on the other hand, is self-consistent in the sense that it computes the white dwarf densities based on the densities of the post-main-sequence stars.

In Fantin et al. (2017), we presented a comparison between the observed white dwarf counts in both the disk and halo and compared them to the resulting simulations. These samples included three selection methods. The first used a colour-colour diagram using photometry from the Next Generation Virgo Cluster Survey (NGVS), the second incorporated ultraviolet data from the GALEX satellite, and the third used proper motions calculated using the NGVS as second epoch observations to the SDSS astrometry. These results are shown in Figure 2.1, where the comparison is made to the Besançon (left-hand column) and TRILEGAL results (right-hand column). We found that the TRILEGAL code over-predicted the number of white dwarfs by a factor of three, particularly the number of young, hot, white dwarfs. Besançon, on the other hand, was consistent with the observations, however, this is expected given that it was calibrated based on similar observations. Furthermore, neither model could reproduce the number of observed halo white dwarfs.

This paper signalled the need for a self-consistent white dwarf population synthesis model which is able to generate mock catalogues that can be used to calculate properties of the Milky Way itself by relating specific input physics to the resulting white dwarf catalogues. This was the motivation to develop a fully self-consistent white dwarf population synthesis code and use it to test various parameters of the Milky Way.

2.2 Model Functions

This section details the construction of a white dwarf population synthesis code. The model incorporates the Milky Way's geometry, as well as important functions used to describe various parameters of both stellar and white dwarf evolution.

2.2.1 Density Profiles

The first step in the model is to populate stars in the Milky Way. This step requires functions describing the stellar density distribution of each Milky Way component. We assume a three-component Galaxy consisting of a thin disk, a thick disk, and a stellar halo. Their density distributions are presented in Table 2.1. We assume a double exponential profile in both Galactocentric radii (R) and height above the plane (z) for the thin and thick disk. We assume a scale length of 2.3 kpc for both

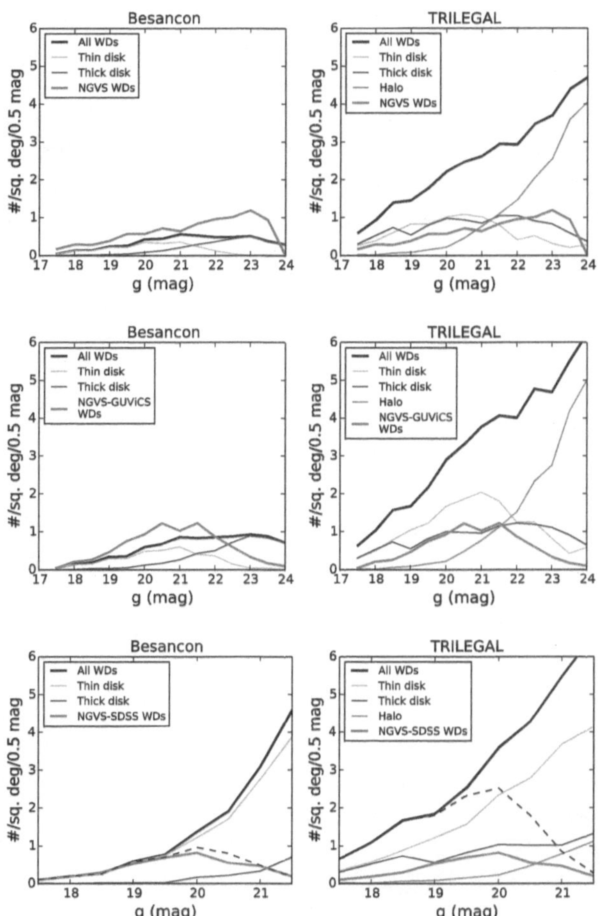

Figure 2.1 Magnitude distributions for three samples of white dwarfs in the NGVS. A sample selected using only the NGVS (top), by combing NGVS with GALEX UV data (middle), and NGVS combined with the SDSS (bottom) are compared to the TRILEGAL (right) and Besançon (left) mock catalogues. The observed WD candidates are shown in red, and the mock WDs are separated into the thin disk (cyan), thick disk (blue) and halo (green) respectively. In the lower panels, the dashed black curves show the mock WD samples after applying corrections to account for the incompleteness suffered by the data. Figure taken from Fantin et al. (2017).

the thin and thick disk, and scale heights of 300 and 800 pc respectively, similar to the mean values presented by Robin et al. (2014).

Stars are spawned at R and z values appropriate for the distributions shown in Table 2.1. The stars are evenly distributed in the angular coordinate, with $0°$ representing the line connecting the solar position in the plane to the Galactic center. The Galactocentric coordinates are then converted to RA and DEC, and a distance from the Sun is calculated assuming the Sun is at a position of $(R, \theta, z) = (8340\,\text{pc}, 0\,\text{pc}, 17\,\text{pc})$ (Reid et al., 2014; Karim & Mamajek, 2017).

2.2.2 Progenitor Masses, Metallicity, and Lifetimes

Now that the positions of the stars are known, the model assigns to each object an initial mass. The mass is randomly sampled from the Kroupa (2001) initial-mass function (IMF), seen in Figure 2.2.

The amount of time it will take each star to become a white dwarf, the progenitor lifetime, is calculated using the analytic stellar lifetimes from Hurley et al. (2000). This functional form takes the initial mass and the metallicity of each star and returns the lifetime of the star (see Figure 2.3). For our model, we assume solar metallicity for the thin disk, $[\text{Fe/H}] = -0.7$ for the thick disk, and $[\text{Fe/H}] = -1.5$ for the halo (Peng et al., 2013).

2.2.3 Star Formation History

Each star is also assigned a birth date (its formation age) that is randomly generated given the functional form for the SFR. Given that the metallicity and kinematics of each population are distinct, we assume each period of star formation can be treated independently (Haywood et al., 2013; Snaith et al., 2014). We assume a skewed Gaussian star formation history for each component,

$$\text{SFR}(t) = \rho_0 \frac{2}{\sigma_t} \phi\left(\frac{t - \xi}{\sigma_t}\right) \Phi\left(\alpha\left(\frac{t - \xi}{\sigma_t}\right)\right) \tag{2.1}$$

where $\phi(t)$ is the standard normal probability distribution that is symmetric about the mean, ξ, and $\Phi(t)$ is its cumulative distribution function. The skewness parameter is α, the standard deviation is given by σ_t, and ρ_0 is the space density. This function allows for four degrees of freedom in order to fit a variety of potential star formation histories.

Table 2.1 Assumed Model Distributions for Each Component

Component	$\rho(R, z)$	$\xi(M)$	SFH	[Fe/H]	$\langle V_\phi \rangle$ (kms^{-1})	σ_U (kms^{-1})	σ_V (kms^{-1})	σ_W (kms^{-1})
Thin disk	$e^{-R/h_R}e^{-z/h_z}$ $h_R = 2300\,\mathrm{pc}$, $h_z = 300\,\mathrm{pc}$	Kroupa	Skewed Gaussian	0.0	−12	33	15	15
Thick disk	$e^{-R/h_R}e^{-z/h_z}$ $h_R = 2300\,\mathrm{pc}$, $h_z = 550\,\mathrm{pc}$	Kroupa	Skewed Gaussian	−0.7	−85	40	32	28
Stellar halo	$r^{-2.44}$	Kroupa	Skewed Gaussian	−1.5	−226	131	106	85

Note. — The Solar position is assumed to be $(R_\odot, \theta_\odot, z_\odot) = (8340\,\mathrm{pc}, 0, 17\,\mathrm{pc})$

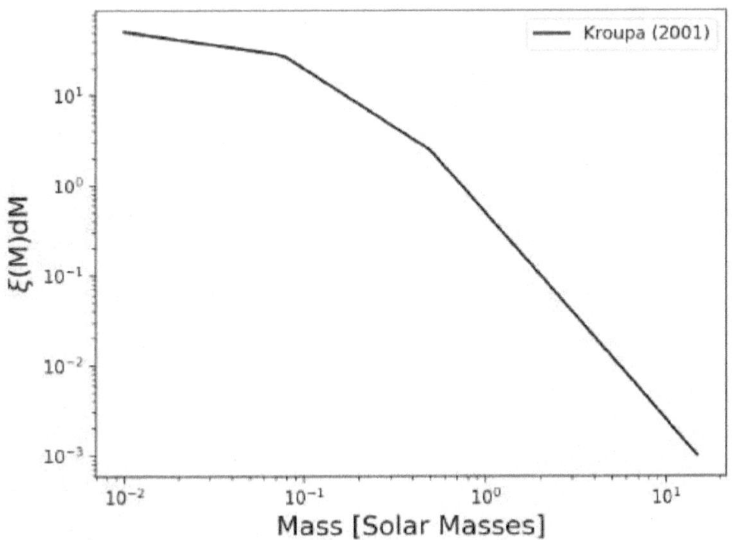

Figure 2.2 The Kroupa (2001) initial-mass function used as part of this model.

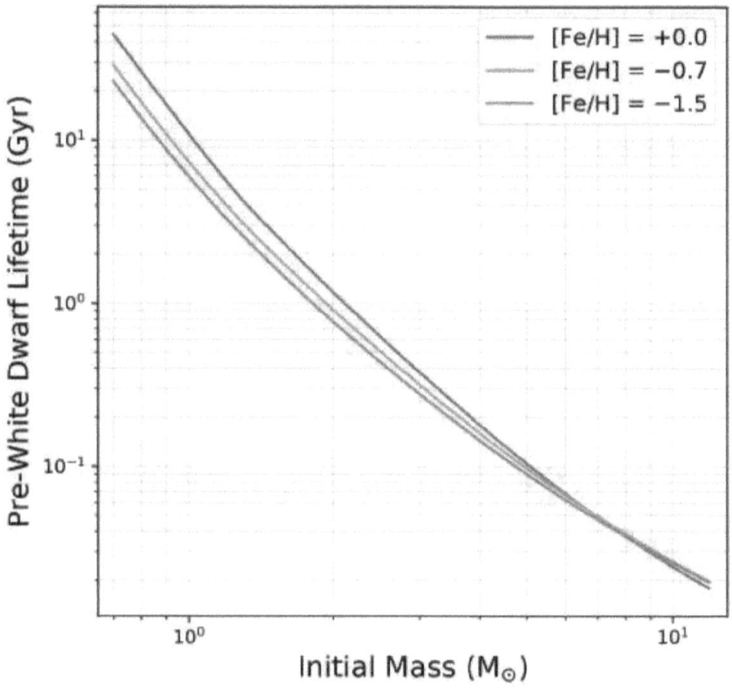

Figure 2.3 Hurley et al. (2000) pre-white dwarf ages as a function of initial mass for three different metallicities representative of the thin disk ([Fe/H] = +0.0), thick disk ([Fe/H] = −0.7), and halo ([Fe/H] = −1.5).

2.2.4 Velocities

A star will have formed a white dwarf if the lifetime up to the end of the asymptotic giant branch phase is less than its formation age, with this difference being the white dwarf cooling time. The white dwarfs are then assigned space velocities according to the distributions presented in Table 2.1. These solar position values were adopted from Robin et al. (2017), who derived them by comparing the Besançon model to kinematic data from RAVE and *Gaia* DR1. These values are representative averages for each component as a small change in velocity will not significantly change the location of an object in an RPMD. These values are similar to those derived by Rowell & Hambly (2011) using white dwarfs in the SuperCOSMOS survey. These velocities are then converted to a proper motion using the equations from Johnson & Soderblom (1987).

2.2.5 The Initial-to-Final Mass Relation

The white dwarf mass, in solar units, is calculated from the initial progenitor mass using the IFMR from Kalirai et al. (2008) or Cummings et al. (2018), both of which are shown in Figure 2.4. The Kalirai et al. (2008) is a single linear function over all masses whereas the Cummings et al. (2018) is a three-component piecewise function.

2.2.6 White Dwarf Synthetic Photometry

The final parameter needed to calculate the photometry of a white dwarf is its spectral type. We include pure H- and pure He-atmosphere white dwarfs in our model. Each white dwarf is designated as one or the other given an input fraction, f_{He}. With the cooling age, white dwarf mass, and atmospheric type in hand, we determine the absolute magnitudes in the required bands using the white dwarf cooling models shown in Figure 1.3. We then determine the reddening in each band at the white dwarfs position using the extinction coefficients and $E(B-V)$ values from the Bayestar 3D dust maps of (Green et al., 2015, 2018) (see Figure 2.5), before converting the model absolute magnitudes to apparent magnitudes. We then add experimental uncertainties based on the observed relation between magnitude and error in the given photometric bands and proper motion. Finally, completeness and selection effects are applied based on the observations.

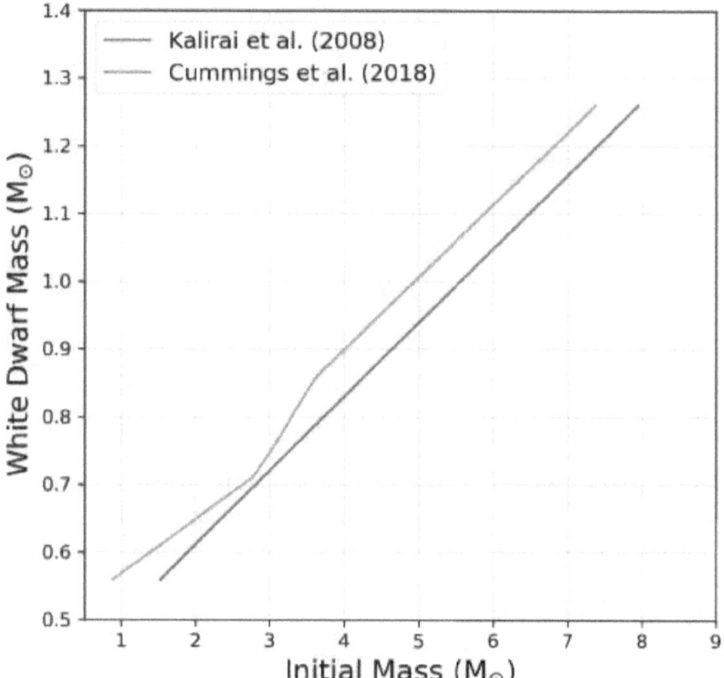

Figure 2.4 The initial-to-final mass relations used by the model.

Figure 2.5 Dust distribution within the Milky Way using the Green et al. (2015) DUSTMAPS model. This map includes the amount of extinction at 100 pc (top), 250 pc (middle), and 500 pc (bottom). Darker colours indicate more dust absorption, showing that there exists very little dust within 100 pc of the Sun.

2.3 Outputs

The model returns a catalogue of simulated white dwarfs within a given footprint. Catalogue information includes positions, distances, masses, cooling ages, spectral types, proper motions, and apparent magnitudes in the given photometric bands.

In Figure 2.6 an example of the resulting equatorial (left) and Galactic (right) positions for the thin disk is shown, with the CFIS-u footprint highlighted in grey. Typical photometric results can be seen in Figure 2.7, which shows colour-colour diagrams (left column) and RPMDs (right column) for young (top row), intermediate-age (middle row) and old populations (bottom row).

While the above description has been tailored for this particular book, we note that the model can readily be adapted for any future study by modifying the band-passes and survey parameters. Examples of this are shown within Chapter 5, where a number of future surveys are simulated.

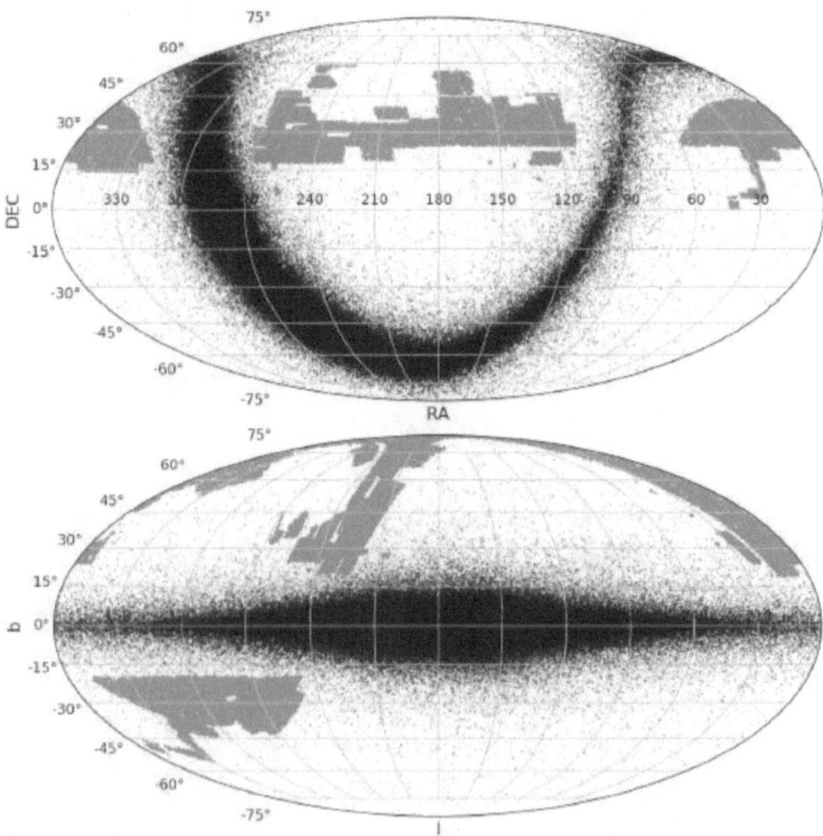

Figure 2.6 *Top:* Equatorial positions of thin disk stars within our model. *Bottom:* The model in Galactic coordinates. In both panels, the shaded region represents the CFIS-u footprint as of the end of the 2018A semester.

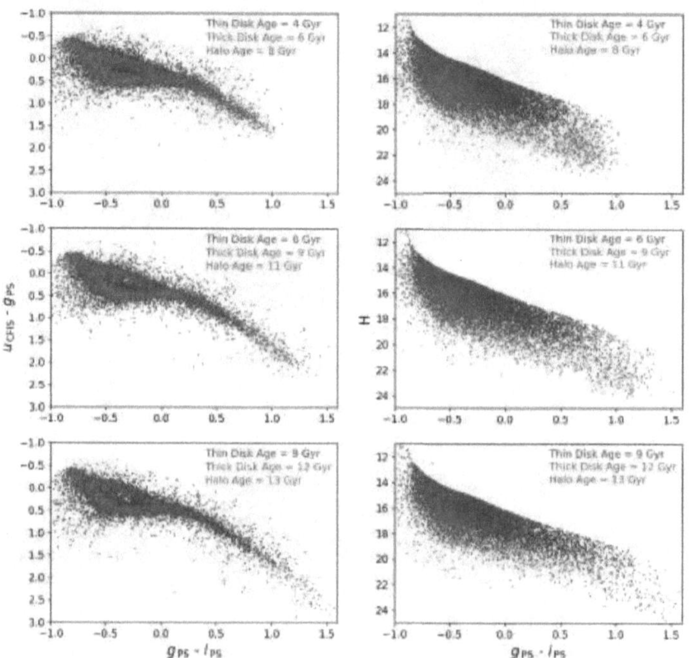

Figure 2.7 Comparing three different realizations of the model with young ages (top row), intermediate ages (middle row), and old ages (bottom row) using colour-colour diagrams (left column) and reduced proper motion diagrams (right column). As the ages of each component increase, a larger number of white dwarfs at redder colours are visible due to the increased cooling ages, resulting in lower temperatures.

Chapter 3

Reconstructing the Milky Way Star Formation History from its White Dwarf Population

3.1 Background

The formation and evolution of disk galaxies, and the Milky Way in particular, has long been an important topic in astronomy since disk galaxies dominate star formation activity in the low-redshift Universe. Previous investigations into the formation and evolution of Milky Way-like galaxies have focused on the assembled mass as a function of redshift (see, e.g, van Dokkum et al., 2013), or compared resolved stellar populations to theoretical isochrones in local galaxies such as Andromeda (Ferguson et al., 2005; Brown et al., 2006). Within the Milky Way itself, much of the attention has been on the stellar metallicity distribution since this, once combined with a model for the gas infall rate, contains information on the star formation rate (SFR) as a function of time (see, e.g, Snaith et al., 2015; Haywood et al., 2018; Toyouchi & Chiba, 2018). Results from GCE models typically show a strong period of initial star formation followed by a slowly decaying trend towards the present day, features usually associated with a thick and thin disk, respectively.

Gilmore & Reid (1983) were the first to present evidence for a possible dichotomy in the properties of the Galactic disk. They used star counts towards the South Galactic Pole and found that the vertical distribution was well described by two separate exponential profiles with different scale heights ("thin" and "thick"). Follow-up studies have shown that the thick disk stars are more alpha-rich, iron-poor, and have a larger velocity dispersion — evidence that they formed much earlier than their thin disk counterparts (Norris et al., 1985; Eggen, 1998). Within the past decade, a renewed emphasis has been placed on the early formation of the Milky Way's disk. The disk dichotomy has been questioned by Bovy et al. (2012) who used mono-abundance populations that represent roughly constant ages to show that each population could be described by a single exponential scale height rather than two global exponential profiles. Haywood et al. (2016), however, argued that the dip in the [α/Fe] distribution seen in APOGEE can only be produced by two separate epochs of star formation, and associates each epoch as belonging to the thin and thick disk respectively (Snaith et al., 2015; Haywood et al., 2018). Haywood et al. (2016) further argues that these two results are consistent if one assumes that the scale height of the thick disk decreases as a function of time so that the differences in structural parameters between the young thick disk and old thin disk are minor. Such a decrease in the scale height, from 800 pc to 340 pc, was found to match star counts in SDSS and 2MASS by Robin et al. (2014), with a mean scale height of 535 $\pm$ 50 pc. Further emphasis has been placed on this transitional epoch with recent results by Helmi et al. (2018) and Belokurov et al. (2018), who show that this epoch of the Milky Way may have played host to a major merger with at least one satellite galaxy, a scenario that would likely play an important role in the formation of the thick disk.

With a complete star formation history, the stellar mass of each component can be estimated. Traditionally, star count models calibrated to the solar neighbourhood have found that the thick disk contributes about 12% of the local mass (Jurić et al., 2008). However, a recent study by Snaith et al. (2015) using the chemical abundances of local F, G, and K-type stars concluded that the thin and thick disk masses may be comparable. This discrepancy likely arises due to the regions of the Milky Way which are probed by each survey. Many models use data confined to the solar neighbourhood, a location at which the thin disk dominates, and thus extrapolating local results to the entire disk becomes problematic.

While the star formation history is imprinted in the chemical enrichment of the Milky Way, stars also leave behind another important tool for astronomers: white dwarfs. With a large enough sample, the complete star formation history of a stellar population can be determined from white dwarfs in two ways. First, with accurate distances one can calculate the luminosity, radius, and effective temperature and convert this into a cooling age (Giammichele et al., 2012). The total age of a white dwarf can be found by adding the cooling age to the progenitor age, which is found by converting the white dwarf mass to a progenitor mass using the initial-to-final mass relation (IFMR). Tremblay et al. (2014) applied this method to the local (20 pc) sample and measured the star formation history in the solar neighbourhood. Their results suggest a roughly constant SFR over the past 5 Gyr, with a slight drop between 5 and 10 Gyr ago (see their Figure 9). This method, however, is typically constrained to small volumes given the distance uncertainties.

The second method is to invert the white dwarf luminosity function (WDLF). Noh & Scalo (1990) showed that varying the functional form for the star formation rate would leave imprints on the resulting WDLF, such as the slope and turn-off. This method was used by Rowell (2013) to invert the SDSS WDLF assembled by Harris et al. (2006), and yielded a resulting star formation history composed of two broad peaks (at 2 and 9 Gyr) separated by a significant dip. The advantage of using this method is that the volume that can be probed is considerably larger; the downside, however, is that volume and completeness corrections must be applied. The recent advances made in wide-field astronomy present an opportunity to increase the sampled volume, and the combination of deep, multi-band photometry and accurate proper motions are ideal for selecting white dwarfs.

In this chapter, we introduce a third technique: forward modelling. This involves simulating a sample of white dwarfs given a star formation prescription. The advantage of forward modelling is that complex systematics, like completeness, observational uncertainties, selection effects, and the structure of the Milky Way can all be accounted for. We use data from the Canada-France Imaging Survey, Pan-STARRS 1, and *Gaia* DR2 to derive the SFH of the Milky Way components using a new population synthesis code. Section 3.2 describes the dataset, while 3.3 describes the fitting method. We present the results in Section 3.4, discuss their implications in Section 3.5, and summarize in Section 3.6.

3.2   Data

The photometric data used in this chapter were acquired as part of the Canada France Imaging Survey (CFIS; Ibata et al., 2017) and the *grizy* Pan-STARRS 1 3π survey (PS1; Chambers et al., 2016). CFIS is an ongoing large-program at the Canada France Hawaii Telescope (CFHT) that aims to obtain 10,000 deg^2 of u- and 5,000 deg^2 of r-band photometry to 5σ point source depths of 24.2 and 24.85 AB mag, respectively. The survey is performed under excellent conditions, with a median u-band image quality of 0.″78. This work makes use of the CFIS-u catalogue, which covers $\sim$4,500 deg^2 as of the end of the 2018A semester. The footprint is highlighted in grey, in both equatorial and Galactic coordinates, as part of Figure 2.6.

The CFIS-u catalogue was merged with PS1 *grizy* forced photometry as described in Thomas et al. (2018). Given that the PS1 and *Gaia* catalogues cover the whole sky as seen from CFHT, our area is set by the CFIS-u footprint. This catalogue retains more than 98% of PS1 detections within the magnitude range of interest. Stars are selected using the recommended star-galaxy classification from Farrow et al. (2014): $i_{\mathrm{PSF}} - i_{\mathrm{Kron}} < 0.05$ mag.

We also merge our catalogue with *Gaia* DR2 (Gaia Collaboration et al., 2018b) in order to obtain proper motions. We omit the photometry acquired by *Gaia* as the photometry from the ground-based surveys is more precise over the magnitude range sampled in this study. We apply the same astrometric quality cuts on the data as presented in Gentile Fusillo et al. (2019),

$$
\begin{aligned}
&\text{ASTROMETRIC\_SIGMA5D\_MAX} < 1.5 \text{ OR} \\
&(\text{ASTROMETRIC\_EXCESS\_NOISE} < 1 \\
&\text{AND PARALLAX\_OVER\_ERROR} > 4 \\
&\text{AND SQRT}(\text{PMRA}^2 + \text{PMDEC}^2) > 10\,\text{mas}),
\end{aligned}
\tag{3.1}
$$

which removes objects with poor astrometric measurements. Specifically, ASTROMETRIC_SIGMA5D_MAX represents the longest axis of the 5-d error ellipsoid, in mas, and large values indicate that one of the astrometric parameters is poorly constrained. ASTROMETRIC_EXCESS_NOISE is a measure of the difference between the best-fit astrometric solution and the data, with large values suggesting a poor astrometric solution. Finally, PARALLAX_OVER_ERROR is the relation between the parallax measurement and its associated error. The limiting magnitude of *Gaia* ($G = 20.7$) represents the primary limitation of our catalogue, and the implications of this are discussed in Section 3.5. Our CFIS-PS1-*Gaia* catalogue contains more than 21.5 million unique point-sources down to the *Gaia* limiting magnitude.

3.2.1 White Dwarf Selection

Using our combined catalogue, we have constructed a reduced proper motion diagram (RPMD) based on photometry from CFIS and PS1 and proper motions from *Gaia* DR2 to separate white dwarfs from other point sources. We omit the parallax measurements as they are not precise to the complete depth of the *Gaia* survey. The reduced proper motion, H (Luyten, 1922), combines the apparent magnitude, m, and the proper motion, μ, and can be used as a proxy for absolute magnitude, M, and tangential velocity, v_t:

$$
\begin{aligned}
H &= m + 5\log\mu + 5 \\
&= M + 5\log v_t - 3.379.
\end{aligned}
\tag{3.2}
$$

Owing to their intrinsic faintness, for a given temperature white dwarfs are well separated in the RPMD from main-sequence stars with similar colours (see, e.g, Harris et al., 2006; Rowell & Hambly, 2011; Fantin et al., 2017; Munn et al., 2017; Fantin et al., 2017). Using model cooling tracks we select white dwarfs with tangential velocities greater than $20\,\mathrm{kms}^{-1}$ as shown in the top panel of Figure 3.1. These synthetic magnitudes have been calculated in the CFIS-*u* and PS1 *grizy* bands for H- and He-atmosphere models using the procedure described in Holberg & Bergeron (2006). The white dwarf cooling sequences are similar to those described in Fontaine et al. (2001) with (50/50) C/O-core compositions, $M_{\mathrm{He}}/M_\star = 10^{-2}$, and $M_{\mathrm{H}}/M_\star = 10^{-4}$ or 10^{-10} for H- and He-atmosphere white dwarfs, respectively [1]. Our selection procedure identifies 25,156 white dwarf candidates within the CFIS-*u* footprint. The resulting sample, along with model H- and He-atmosphere tracks, can be seen in colour-colour space in the bottom panel of Figure 3.1.

Previous selections using the RPMD have obtained contamination fractions from non-white dwarfs (typically hot sdO, sdB, or cooler main-sequence stars) of a few

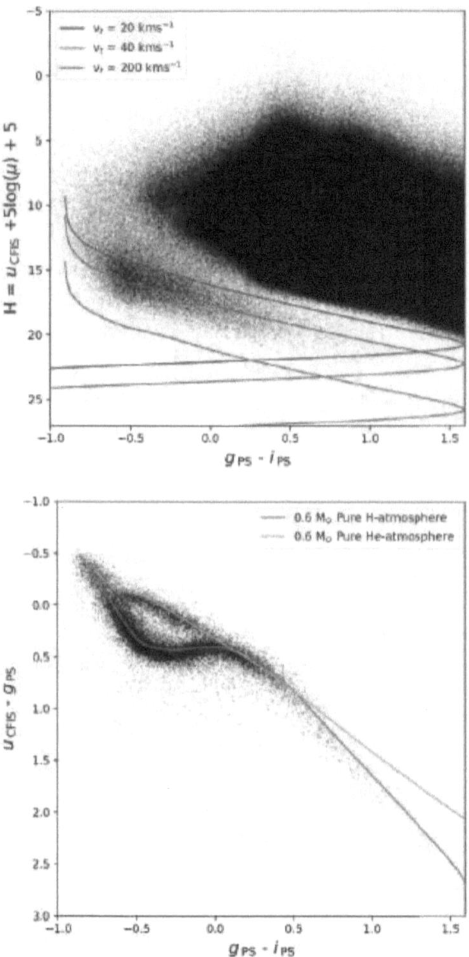

Figure 3.1 *Top:* CFIS-PS1-*Gaia* reduced proper motion diagram (RPMD) showing all 24.5 million sources. White dwarfs for this study were selected if they had inferred tangential velocities greater than $20\,\mathrm{kms}^{-1}$ (the region bound by the blue line) *Bottom:* Colour-colour diagram showing the resulting white dwarf candidates selected from the RPMD. For reference, $0.6\,\mathrm{M}_\odot$ H- and He-atmosphere model tracks are plotted in red and cyan, respectively.

percent (e.g, Harris et al., 2006) at a velocity cut of $v_t > 30\,\mathrm{kms}^{-1}$, with the rate increasing for lower velocity cuts (Kilic et al., 2010). Given that *Gaia* DR2 has much better astrometry than previous studies, we have chosen to lower our velocity cut to $v_t > 20\,\mathrm{kms}^{-1}$. Matching our objects to SIMBAD, we find 3688 spectroscopic matches, of which 3589 are white dwarfs. The remaining objects are a mix of hot subdwarfs (69 objects), QSOs with spurious proper motion measurements (20 objects), and 10 miscellaneous objects including main-sequence stars. The resulting contamination rate is thus at least $\sim3\%$, a value that has been incorporated into the results presented in Section 3.4.

The colour-colour diagram highlights the precise photometry of CFIS and PS1 and the ability to separate DA (H-lines dominant) and DB (He-lines dominant) white dwarfs between $-0.6 < g_{PS} - i_{PS} < 0.0$ mag. The RPMD also highlights the precise astrometry of *Gaia* DR2, which results in a clean sequence of white dwarfs in colour-colour space. In Figure 3.2 we present the observational uncertainties in both the photometry and astrometry as a function of magnitude for our white dwarf sample, highlighting the depth of the CFIS-u band and the precision of the *Gaia* DR2 proper motions.

3.2.2 Completeness

We calculate the completeness within our dataset by comparing our CFIS-PS1-*Gaia* catalogue to the CFIS-PS1 catalogue. Thomas et al. (2018) showed the PS1 g-band is complete to 22.0 mag, which is deeper than our sample (see Figure 3.2) and thus we assume that the CFIS-PS1 dataset is complete over the magnitude range covered by the *Gaia* survey. Our completeness curves for the PS1 g- and i-bands, as well as the CFIS-u band, can be seen in Figure 3.3.

3.3 Fitting Method: Approximate Bayesian Computation MCMC

The model presented in Chapter 2 is used to compare to the data. We begin by parameterizing the SFH for each Galactic component following equation 2.1. This results in four parameters for each component: the mean, ξ, which we will call the functional age, the scale, σ_t, the skewness α, and the star formation rate. This results in a total of 12 parameters. We also fit for the ratio between the hydrogen and helium atmosphere white dwarfs, which we call the He fraction, f_{He}. This fraction is assumed

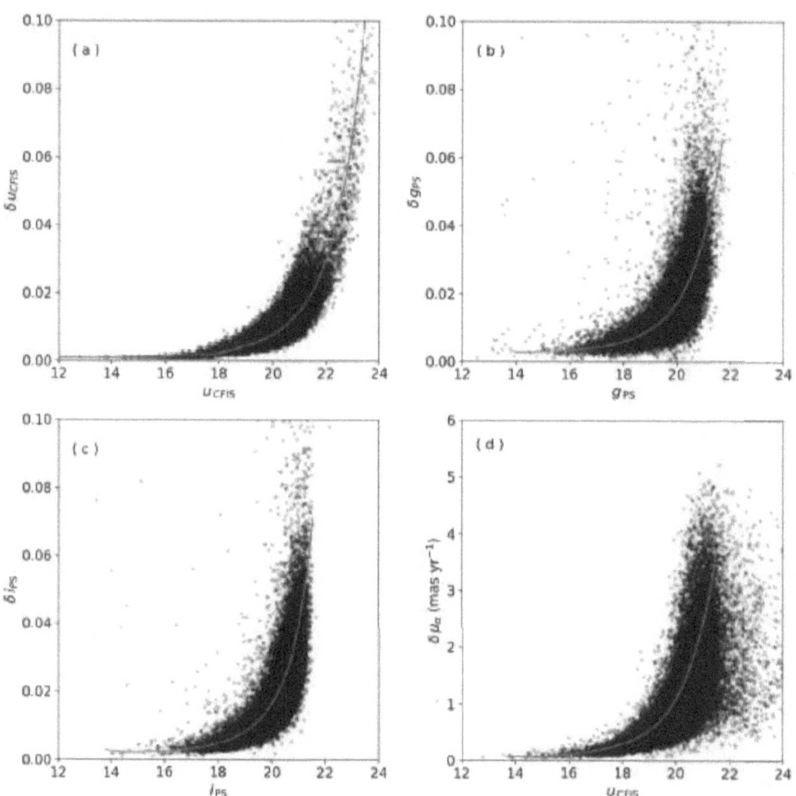

Figure 3.2 Observational uncertainties in (a) CFIS u, (b) PS1 g, (c) PS1 i, and (d) proper motions as a function of magnitude in our white dwarf sample. The model, as described in Section 5.3, samples a Gaussian at each point with a mean (given by the red line showing a polynomial fit) and standard deviation in order to obtain an uncertainty value for each mock white dwarf.

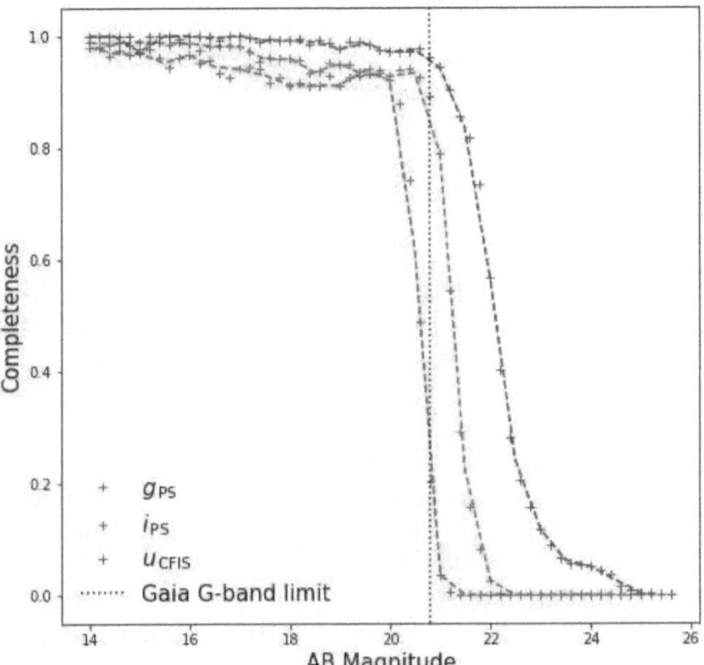

Figure 3.3 The resulting completeness as a function of magnitude for our CFIS-PS1-*Gaia* sample in the CFIS u- (blue), PS1 g- (green), and PS1 i-band (red) respectively. These completeness functions were calculated assuming that the CFIS-PS1 catalogue is complete over the magnitude range of *Gaia* DR2, which was shown to be true in Thomas et al. (2018).

to be equal for all three populations. This brings the total number of parameters to 13.

We integrate a given SFH to determine the total stellar mass and select the corresponding number of stars contained in each Milky Way component. For each star, we then generate a formation age given the SFH. We then remove every star that does not fall within the CFIS footprint and/or will not have formed a white dwarf at the present day. The cooling age for those stars that remain is then calculated as being the difference between the formation age and the progenitor lifetime.

With the cooling ages, white dwarf masses, and spectral type, we then calculate apparent magnitudes in the CFIS and PS1 bands using the model cooling curves described in Section 3.2. Observational uncertainties are added given the relations shown in Figure 3.2. A photometric completeness correction is then made given the curves presented in Figure 3.3.

Finally, we perform the same selection methods on the model as we do on the data. We begin by applying identical distributions in the *Gaia* parameters included in equation 3.1 to our model white dwarfs before making the same selection as presented in equation 3.1. Using the proper motion we calculate the reduced proper motion using equation 3.2 and select those stars whose tangential velocity is greater than the $20\,\mathrm{kms}^{-1}$ curve (see Figure 3.1).

We use *astroABC* (Jennings & Madigan, 2017), a python based approximate Bayesian computation (ABC) Markov chain Monte Carlo (MCMC) approach in order to calibrate our model. This so-called 'likelihood free' method is useful in cases where a likelihood function is difficult to calculate or unknown altogether. Unlike traditional MCMC methods used in astronomy (see, e.g. *emcee*; Foreman-Mackey et al., 2013), ABC MCMC does not calculate an explicit likelihood but instead relies on forward modelling. Given a set of model parameters, a simulated dataset is produced and compared to the data in order to either accept or reject the proposed parameters. More specifically, if we draw parameters θ, we accept these parameters if $\rho(D - D^*(\theta)) < \epsilon$, where ρ is the distance metric between the dataset, D, and the simulated data given θ, $D^*(\theta)$, and ϵ is called the tolerance threshold. The distance metric used to compare the data and the model is

$$\rho(D_i - D_i^*(\theta)) = \sum_i \frac{(D_i - D_i^*(\theta))^2}{2\sigma_i^2}. \tag{3.3}$$

The result at each step is a set of independent samples below a generated tolerance threshold (Marjoram et al., 2003).

The algorithm starts off with a large threshold and shrinks it at each step in order to approximate the probability density function (PDF). When this threshold is exactly zero, then one is sampling directly from the posterior PDF; however, in that case, the acceptance fraction declines sharply as it is difficult to perfectly simulate a dataset. On the other hand, if ϵ is too large, then the algorithm is sampling from the prior. Thus, the choice of ϵ determines how much of an approximation the resulting PDFs are while minimizing the required total computational time (Marjoram et al., 2003).

As described in Marjoram et al. (2003), ABC methods typically require lower-dimensional summary statistics, $S(D)$, which are representative of the whole dataset in order to reduce computational costs. This summary statistic must encompass any required information contained within the data itself without neglecting anything important. Similar to Mor et al. (2018), we use a combination of binned colours, magnitudes, and proper motions. Specifically, we compare our real and simulated data using a 3-D histogram composed of a colour-colour diagram (g-i, u-g), and the reduced proper motion, H. This combination of colours and proper motions was chosen in order to encompass both age and SFR information (colours), as well as the contributions from each component (apparent magnitudes and proper motions via H). Furthermore, the colour-colour diagram contains information regarding the He fraction in our sample, since they are well separated due to the exquisite u-band photometry provided by CFIS (see Figure 3.1) Thus, in equation 3.3, i represents each bin in the histogram, and σ is the sum in quadrature of the uncertainty in both the model and data in each bin. For instances where the model predicts a star, but we do not observe one, we use substitute D_i+1 into equation 3.3 in order to penalize the model by increasing the distance. We then run *astroABC* with 1000 particles to obtain the approximate PDFs for the age, duration, and SFR for the three Galactic components, as well as the He fraction. We use uniform priors for every variable, with the only constraint being that the age of the components cannot exceed the age of the Universe, which we set to $13.8\,\mathrm{Gyr}$ (Planck Collaboration et al., 2018), and that the scale parameters, σ_t, must be greater than zero.

3.4 Results and comparison to the literature

The resulting corner plot containing the PDFs can be seen in Figure 3.4. The results from our fit can be seen in the bottom panels of Figure 3.5, which show a model instance with the mean parameters as estimated by the final step of the Markov chain (the central line in the histograms from Figure 3.4). The top two panels show the actual data for comparison. We are able to reproduce the thin and thick disk population quite well, although the halo parameters are much less constrained. This is a result of the *Gaia* limiting magnitude since the intrinsic faintness and scarcity of halo white dwarfs results in very few appearing in the dataset. For reference, only 44 objects lie below the 200 kms^{-1} model track in Figure 3.1, which is typically associated with the Galactic halo. This sample is smaller than that obtained with deeper SDSS photometry by (Munn et al., 2017), which was used by Kilic et al. (2017) to obtain a halo age of $12.5^{+1.4}_{-3.4}$ Gyr. A larger sample will be required to better constrain the halo parameters — in particular, a sample that includes objects beyond the turnover in the halo luminosity function. This will require deeper proper motion surveys, such as LSST, which will be discussed in Chapter 5. Within the next decade, upcoming imaging surveys such as the Roman Space Telescope (Spergel et al., 2015) and LSST (LSST Science Collaboration et al., 2009) will provide proper motions to unprecedented depths, allowing for a more comprehensive study of the Galactic halo.

3.4.1 Our Sample, in the Context of the Milky Way

In order to properly compare with previous results, we begin by showing the volume that is sampled by our model. The resulting sight-lines in both R (the distance from the centre of the Milky Way) and z (the distance from the mid-plane) can be seen in the left-hand panel of Figure 3.6. The cyan dashed lines show heliocentric distances of 1, 2, and 3 kpc, respectively, showing that the vast majority of our sample lies within one kiloparsec of the Sun. This is confirmed in the right-hand panel which shows the distance distributions of our model white dwarfs. The result is that only 4% of white dwarfs in our sample lie at distances greater than 1 kpc, with a median distance of 388 pc. Given these values, we consider our sample to be representative of the solar neighbourhood and not necessarily the whole disk.

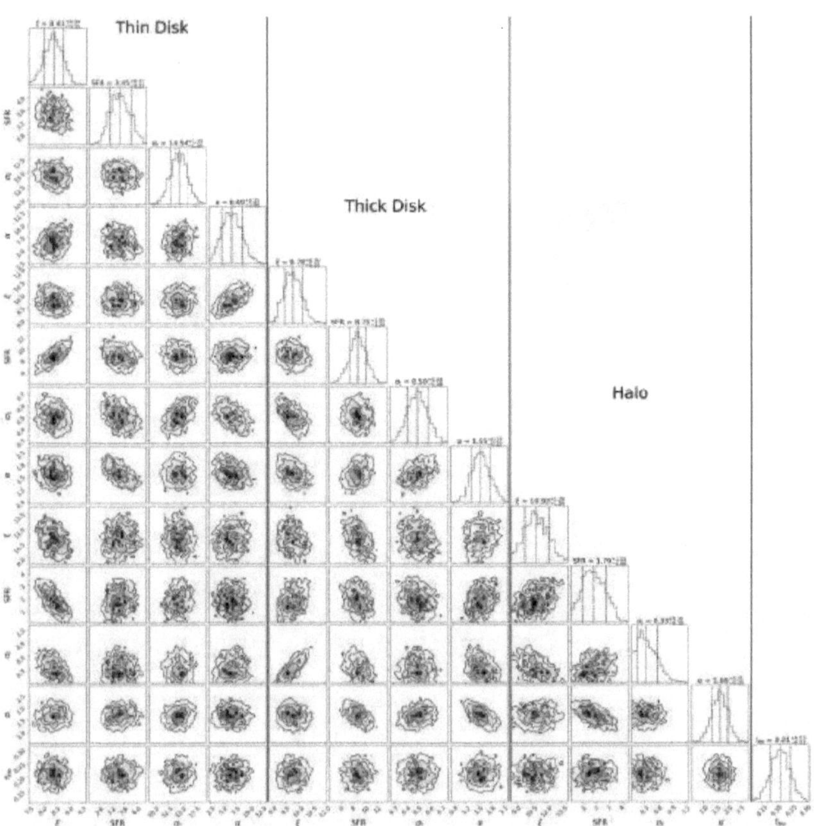

Figure 3.4 Posterior distributions for the 13-dimensional parameter space sampled with *astroABC*. From left to right we show the mean functional age (ξ, Gyr), star formation rate (SFR, $M_\odot \mathrm{yr}^{-1}$), standard deviation (σ_t), and skew (α) for the thin disk, thick disk, and halo. See equation 2.1 for the definition of each parameter. The final histogram shows the fraction of white dwarfs with helium atmospheres, f_{He}.

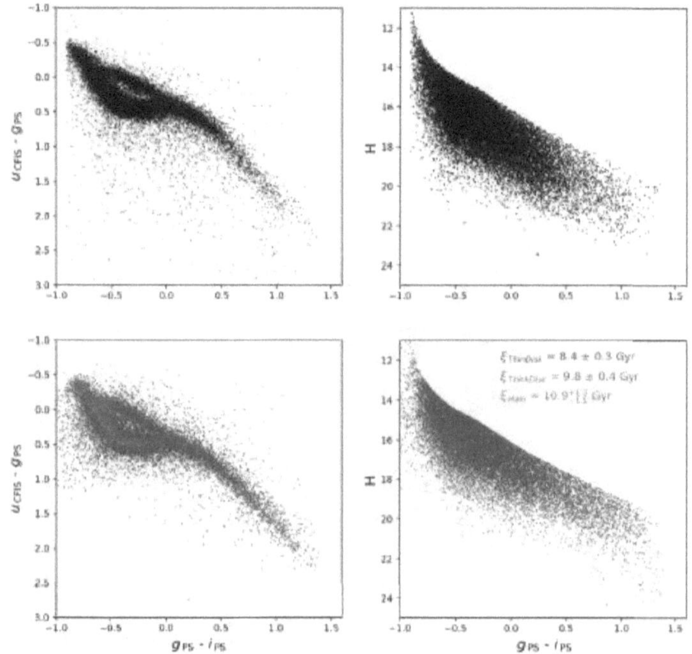

Figure 3.5 Comparing our best fit model (bottom panels) to the data (top panels) in both colour-colour (left column) and RPMD (right column). The best fit ages are shows in the bottom right panel.

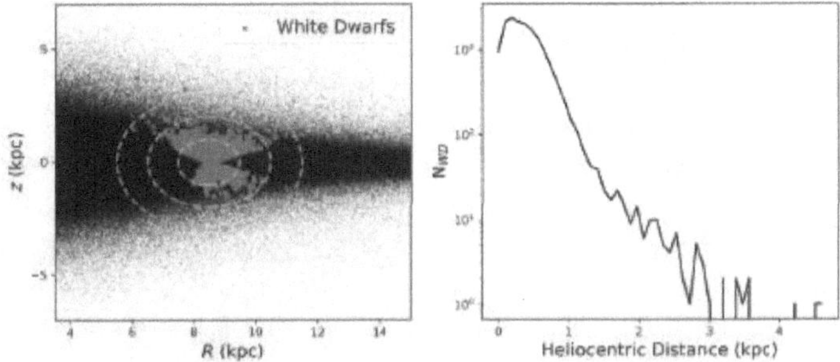

Figure 3.6 *Left:* Positions of our model white dwarfs (red) within the Milky Way (black). The concentric cyan rings represent heliocentric distances of 1, 2, and 3 kpc respectively. *Right:* Heliocentric distance distributions of our model white dwarfs. Our sample consists mainly of local white dwarfs, with 96% having a distance less than 1 kpc.

3.4.2 Star Formation History

Our resulting SFH can be seen in the top panel of Figure 3.7, where we have sampled the resulting PDFs and plotted the mean (black) and 1σ contours (grey). Our results show an initial period of thick disk star formation, lasting 3 Gyr and peaking at (9.8 ± 0.3) Gyr. This is followed by a nearly 1 Gyr decline in star formation before finishing with a nearly constant SFR for the thin disk at approximately $3.5\,\mathrm{M_\odot yr^{-1}}$. Below, we present a comparison between our SFH and those found in the literature. We make note of a caveat to these comparisons, which is that we assume that each component is defined by a unimodal SFH. Thus, if a given component contains multiple bursts, we will fit the average through the bursts (see Section 3.5 for further discussion).

The star formation and, in turn, the mass assembly, history of the Milky Way has been an open question for decades. With the increase in accurate distances in the Solar neighbourhood from surveys like *Hipparcos* (Hip, 1997), the local SFH could be derived by comparing the synthetic colour-magnitude diagrams (CMD) to observations. This process was performed by Vergely et al. (2002) using *Hipparcos* stars brighter than $V = 8$, as well as Cignoni et al. (2006) who used all *Hipparcos* stars within 80 pc. Both of their samples are dominated by thin disk objects given their selection criteria, and in turn, they show a slowly rising star formation until it hits a peak 2 Gyr ago. These SFHs can be seen in the top panel of Figure 3.8 as the red dashed line and dotted cyan line respectively. In this figure, we have scaled our star formation history and hence only the shapes should be compared. We note that neither SFH recovers a strong peak early-on in the formation of the Milky Way (the Cignoni et al. (2006) result has a slight uptick in the $10-12$ Gyr bin) and this may be a result of their samples being dominated by younger thin disk objects.

We also compare our SFH to that obtained by Rowell (2013) who implemented an algorithm to invert the WDLF in order to derive the local star formation history. The resulting SFH, computed from the Harris et al. (2006) WDLF, is shown as the dot-dashed green line in the top panel of Figure 3.8. Both SFHs are normalized, and given the varying units obtained by each study only a comparison of the shape is appropriate. Both SFHs show two stages of star formation, however, we do not see a recent peak $2-3$ Gyr ago, and the peak in our SFR attributed to the thick disk happens about 2 Gyr earlier. This is likely due to the larger volume sampled by our CFIS-PS1-*Gaia* sample, which in turn will allow us to better constrain this time-frame since our sample will contain a larger fraction of thick disk objects. It is worth

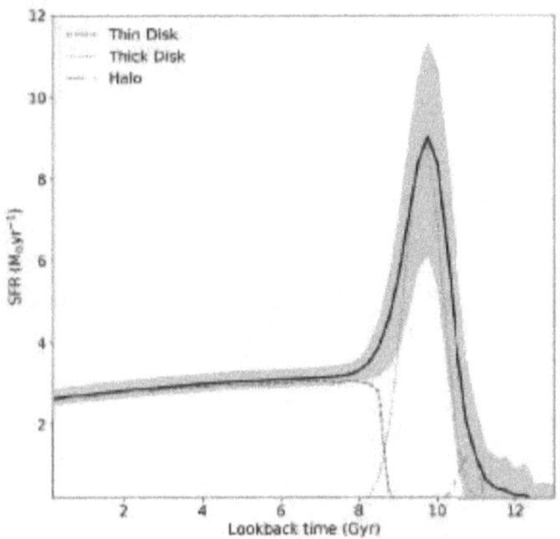

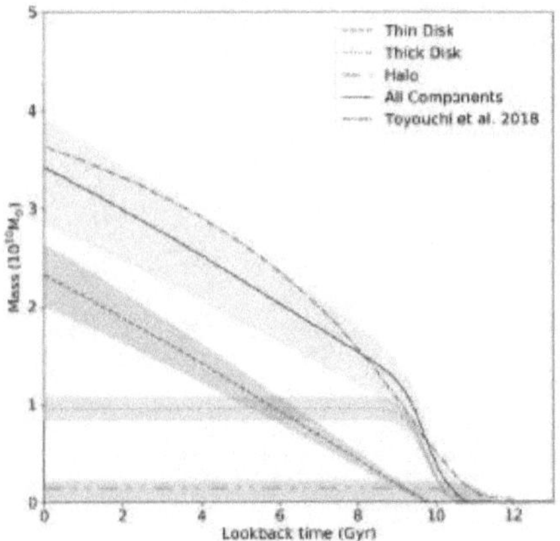

Figure 3.7 *Top*: Milky Way star formation rate as a function of lookback time. *Bottom*: The cumulative mass as a function of lookback time with the contribution from the thin disk (dashed blue), thick disk (dotted green) and halo (double dot-dash red) highlighted.

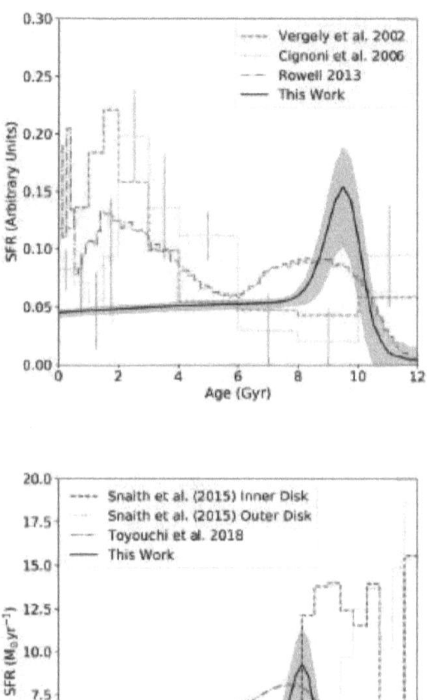

Figure 3.8 *Left:* Our star formation history (black) compared to *Hipparcos* results by Vergely et al. (2002) (dashed-red) and Cignoni et al. (2006) (dotted-cyan). Also shown is the result of Rowell (2013), who inverted the white dwarf luminosity function of Harris et al. (2006) (green). *Right:* Comparing our star formation history to results obtained via Galactic chemical evolution models from Snaith et al. (2015) and Toyouchi & Chiba (2018). Due to the varying units presented by each study, the SFHs have been normalized, and therefore only the shape should be compared.

noting that similarly a burst in the SFR at 600 Myr is seen in the luminosity function of the local 40 pc white dwarf sample from Torres & García-Berro (2016), but not in the deeper luminosity function generated by Munn et al. (2017). Kilic et al. (2017) concluded that this discrepancy is a result of the local sample being confined to the mid-plane of the Milky Way, where most recent star formation occurs.

In the bottom panel of Figure 3.8, we compare our results to other SFHs found in the literature through GCE models. Toyouchi & Chiba (2018) used an open-box model in order to reproduce the radial metallicity distribution obtained using APOGEE data in Hayden et al. (2015). This result was obtained using a sample of red-giant stars with $3 < R < 15$ kpc and $|z| < 2$ kpc, which is a much larger volume than our sample. Their result shows a rise in star formation beginning at 12 Gyr with a peak at 9 Gyr. This is followed by a roughly constant SFR to the present day. Our SFH matches the Toyouchi & Chiba (2018) results quite well at recent and ancient times but differs slightly at intermediate ages.

Also plotted are results by Snaith et al. (2015) who compared their GCE model to a sample of abundances acquired as part of the HARPS survey. They model the Milky Way disk with two components: an inner ($R < 7$ kpc) closed-box and an outer ($R > 9$ kpc) component which can accrete gas. Their resulting inner disk SFH shows a two-phase formation scenario separated by a period of inactivity lasting roughly one Gyr. A follow-up paper by Haywood et al. (2018) showed that the inner disk model is compatible with APOGEE data, and argued that the gap could be caused by the formation of a bar.

Their inner and outer disk results have been plotted in purple and orange, respectively, in Figure 3.8. Given the relative nature of their presented star formation history, we have scaled their SFH so that both periods associated with the thin disk are equal. The inner disk shows an initial burst of star formation, which they attribute to the thick disk, and ends with a relatively constant SFR over the past 7 Gyr associated with the thin disk. Comparing the general shape of their inner disk SFH, an initial peak of SFH followed by a decline to a nearly constant SFR, shows that the two models are consistent, however, we do not *require* a gap in our SFH. This is likely because this is a feature of the inner disk, and since the gap is thought to be a result of the formation of a bar, it did not have a large effect on the local star formation we measure.

Our results differ in terms of the SFH of the thick disk, particularly at the earliest stages of the Milky Way's formation. This is likely a consequence of the large uncertainties in both models at the earliest epochs of the Milky Way and in particular the difficulty of detecting white dwarfs with cooling ages exceeding 11 Gyr. This could also be a result of our choice of scale height (See section 3.5), as an increased scale height at these epochs would increase our SFR.

Haywood et al. (2019) argue that the solar neighbourhood is better described by the outer disk evolution, given the enrichment history of the outer disk. In their outer disk scenario, their simulation includes a single accretion event at 10 Gyr (and thus the SFR between 10 and 13.8 Gyr is unconstrained) and results in a consistent SFR over the past 10 Gyr, similar to the inner disk. This is roughly consistent with our results, particularly within the past 8 Gyr.

3.4.3 Component Masses

The bottom panel of Figure 3.7 shows the cumulative mass of the Milky Way as a function of time, again with all 1000 particles shown in grey. This is done by integrating our SFH and density functions over the entire Milky Way. We measure a total stellar mass of $(3.4 \pm 0.6) \times 10^{10}\,\mathrm{M_\odot}$ for our thin disk, thick disk, and stellar halo. Also plotted is the result of Toyouchi & Chiba (2018) which is consistent within uncertainties with our result. Given the relative nature of the other star formation histories, we have chosen not to make a direct comparison with the results plotted in Figure 3.8.

The mass of the Milky Way's stellar disk is an important astronomical property, and as such, has been extensively studied. Our result is lower than many previous results. Recent mass models of the Milky Way have found a total disk mass between $3.6 - 5.5 \times 10^{10}\,\mathrm{M_\odot}$ (see, e.g, Gerhard, 2002; Flynn et al., 2006; McMillan, 2011; Bovy & Rix, 2013). It should be noted, however, that these estimates are highly dependent on the model assumptions. These assumptions include the scale length and scale height, IMF, and whether or not a bulge/bar is included. Furthermore, it should be reiterated that our survey volume is rather small, and thus extrapolating our volume to the entire disk will inevitably contain a systematic uncertainty.

3.4.4 Local Stellar Density and White Dwarf Number Density

Given the systematic uncertainties associated with extrapolating our result to the entire disk, a more appropriate comparison is likely with the local stellar density. The stellar density as calculated by our model is $(0.036 \pm 0.004)\ \mathrm{M_\odot pc^{-3}}$ for the thin disk, $(4.5 \pm 0.5) \times 10^{-3}$ $M_\odot pc^{-3}$ for the thick disk, and $(3.2 \pm 3.1) \times 10^{-6}$ $M_\odot pc^{-3}$ for the halo. Our values agree with Bovy (2017), who used *Gaia* DR1 to calculate a main-sequence stellar density of 0.040 ± 0.002 $M_\odot pc^{-3}$.

The resulting white dwarf densities are $(4.8 \pm 0.4) \times 10^{-3} pc^{-3}$ for the thin disk, $(1.0 \pm 0.2) \times 10^{-3} pc^{-3}$ for the thick disk, and $(6.3 \pm 2.4) \times 10^{-6} pc^{-3}$ for the halo. This corresponds to a total white dwarf number density of $(5.8 \pm 0.5) \times 10^{-3} pc^{-3}$. Our values are consistent with the estimates of $(5.5 \pm 0.1) \times 10^{-3} pc^{-3}$ from Munn et al. (2017), and marginally higher than results of $4.6 \times 10^{-3} pc^{-3}$ from Harris et al. (2006), $(4.8 \pm 0.4) \times 10^{-3} pc^{-3}$ from Torres et al. (2019), and $(2.81 \pm 0.52) \times 10^{-3} pc^{-3}$ from Fantin et al. (2017).

Our resulting white dwarf contributions within the solar neighbourhood break down to $(83 \pm 5)\%$, $(17 \pm 3)\%$, and $(0.11 \pm 0.05)\%$ from the thin disk, thick disk, and stellar halo respectively. Rowell & Hambly (2011) constructed thin disk, thick disk, and halo luminosity functions using a statistical approach based on kinematics and photometry from the SuperCOSMOS survey in order to obtain local fractions of 79, 16, and 5 percent, respectively. This result was consistent with Reid (2005) who found that thick disk white dwarfs should comprise roughly 20% of the local population. Our thick disk fraction is marginally lower than the recent result by Torres et al. (2019) who obtained fractional contributions of 74, 25, 1 % from the thin disk, thick disk, and halo respectively.

3.4.5 Component Ages

The white dwarf luminosity function has been used as a tool to calculate the ages of various populations within the Milky Way, including its field population. Our functional ages, ξ (see Equation 2.1), for the thin disk and thick disk are $8.41^{+0.34}_{-0.35}$ Gyr and $9.78^{+0.36}_{-0.33}$ Gyr, respectively. Since the age can be defined as the onset of star formation, we also present these values as (8.5 ± 0.3) Gyr for the thin disk and (11.3 ± 0.3) Gyr for the thick disk.

The first age measurement using white dwarfs was preformed by Winget et al. (1987) who obtained an age of (9.3 ± 2.0) Gyr for the Galactic disk. Following this, Leggett et al. (1998) used spectroscopic data from a sample of 43 white dwarfs to obtain an age of (8 ± 1.5) Gyr for the disk. Both of these samples were likely dominated by thin disk objects, and this is reflected in their age measurements.

Our SFH is consistent with those derived from the oldest open clusters, which are typically associated with the thin disk. Bedin et al. (2005) and García-Berro et al. (2010) used the WDLF to determine an age of $\sim$8 Gyr for NGC 6791, an open cluster with solar-like metallicity. This cluster likely formed at the onset of thin disk star formation given its age and metallicity.

Our thick disk age is consistent with metal-rich globular clusters, like 47 Tucanae, which have metallicities comparable to thick disk field stars ([Fe/H] $= -0.75$). Hansen et al. (2013) used *Hubble Space Telescope* photometry to observe the WDLF and obtain an age of (9.9 ± 0.7) Gyr. In our star formation history, this would place its formation near the peak of the thick disk star formation period.

Kilic et al. (2017) used the WDLF presented by Munn et al. (2017), who performed second epoch follow-up photometry within the SDSS footprint to calculate proper motions and select their white dwarf sample. The ages of our thin disk and thick disk are consistent within the errors with their results of $7.4-8.2$ Gyr and $9.5-9.9$ Gyr.

Although less constrained than our disk ages, our resulting functional halo age is $10.92^{+1.3}_{-1.1}$ Gyr. This translates into an onset age of (12.3 ± 1.3) Gyr. This result is consistent with metal-poor globular clusters associated with the stellar halo, which typically have ages of $11-13$ Gyr. For example, Hansen et al. (2013) obtained an age of (11.7 ± 0.3) Gyr for NGC 6397 and Bedin et al. (2009) found an age of (11.6 ± 0.6) Gyr for M4, both using the WDLF. Our result is also consistent with Kilic et al. (2017), who presented an age of $12.5^{+1.4}_{-3.4}$ Gyr.

3.4.6 He Fraction

The resulting He fraction in our sample is $(21 \pm 3)\%$. This is consistent with Bergeron et al. (2011), who fit model atmospheres to a spectroscopic white dwarf sample from the Palomar Green Survey. They showed that approximately 20% of white dwarfs with temperatures below 20,000 K appeared to contain a helium atmosphere. This result was followed up by Genest-Beaulieu & Bergeron (2019), who used SDSS spectra to show that the fraction of DB stars increases at lower temperatures (see their Figure 23). This fraction was found to be approximately 5% at temperatures greater than 20,000 K, with an increase to $20-25\%$ at 12,000 K. Using a sample of local white dwarfs Limoges et al. (2015) also showed that 25% of their sample was best described by models containing a helium atmosphere.

Our measurement is also consistent with recent results from Kilic et al. (2018) who used a colour-magnitude diagram, generated using SDSS photometry and *Gaia*

DR2 astrometry, to obtain a DB fraction of 36 ± 2% for a sample of white dwarfs within 100 pc. Since He-atmosphere white dwarfs are most pronounced at higher temperatures their result was quoted within a region of colour-magnitude space encompassing the bifurcation of DA and DB white dwarfs due to the increased strength of the Balmer lines. In the temperature region, He-atmosphere white dwarfs are classified as type DB since He I lines are present in their spectra. Applying the same selection area, $M_{u_{\mathrm{SDSS}}}$ between 10 and 14 and u_{SDSS} - g_{SDSS} between -0.4 and $+0.6$, to our CFIS and PS1 model colour-magnitude diagram produces a consistent DB fraction of 34 ± 3%.

3.5 Discussion

As Figure 3.7 shows, our results suggest the star formation of the Milky Way disk began (11.3 ± 0.3) Gyr ago. Following a brief period of halo formation, the thick disk rapidly begins to form stars until it reaches a peak (9.8 ± 0.3) Gyr ago. This period is followed by a decline in thick disk star formation. The thin disk then begins to form stars (8.5 ± 0.4) Gyr ago.

The transition between the thin and thick disk SFHs was a fairly active period in the Milky Way's history. Recent results by Helmi et al. (2018) and Belokurov et al. (2018) provide evidence that this epoch was dominated by the accretion of at least one satellite galaxy, and may have contributed to the rapid transition between the thick and thin disk. The resulting merger would have heated up the existing stars to form a thick disk approximately 10 Gyr ago (Helmi et al., 2018). Given that a merger can enhance the SFR for a short period before quenching (Di Matteo et al., 2008), our results suggest that the merger occurred shortly before our peak SFR at (9.8 ± 0.3) Gyr, and before the onset of star formation in the thin disk. Thus, this scenario is consistent with a merger triggering the epoch of peak star formation in the Milky Way and the start of star formation in what we now call the thin disk.

In this section, we discuss the impact of our assumptions on these results, including our choice of star formation prescription, and discuss the next steps for this study.

3.5.1 Effect of Star Formation Prescription

As shown in equation 2.1, we assume skewed Gaussian functions for the three components, as allowing a skewness increases the degrees of freedom of the function. In this sense, we have assumed unimodal functions for the star formation histories, and we will inevitably fit over any potential multimodal peaks in the star formation. Specifically, if the star formation history of a single component is composed of multiple bursts, then we will fit the average star formation throughout these periods of increased and decreased star formation. To examine this possibility, we have plotted the fractional residual between the data and the best fit model in Figure 3.9.

Figure 3.9 shows a deficit of white dwarfs within the solid black box and an excess of white dwarfs within the dashed box. These areas represent mean formation ages of (3.3 ± 1.8) Gyr and (5.8 ± 1.1) Gyr respectively. This suggests that we are underestimating the star formation rate around 3 Gyr and overestimating at 6 Gyr by roughly 50% and 30%, respectively. This is consistent with the results of Rowell (2013) and Vergely et al. (2002), as shown in the top panel of Figure 3.8, who see an increase in star formation between 2 and 3 Gyr and a resulting decrease in the vicinity of 6 Gyr. This result is also in line with the recent work by Mor et al. (2019), who compared simulations of the Besançon Galactic model with the colours, magnitudes, and parallaxes from *Gaia* DR2, which shows a burst of star formation between 2 and 3 Gyr in the past. The slight bimodality suggested by Figure 3.9 is, however, not seen in the GCE models of Snaith et al. (2014) and Toyouchi & Chiba (2018), as seen in the bottom panel of Figure 3.8.

3.5.2 Effect of the Scale Height

In our model, we have assumed a three-component Galaxy with constant scale heights. This assumption will inevitably impact our SFH, as a decrease in the scale height will result in a lower star formation rate. This is because more stars will reside closer to the plane, and will have a higher probability of passing our observational criteria.

Robin et al. (2014) found that the scale height for the thick disk (described using a sech^2 distribution) likely contracted from 800 pc to 350 pc between 12 and 10 Gyr. Given this scenario, our SFH would under-predict the SFR while the scale height was greater than 550 pc and over-predict while it was below this value. However, given the relatively small distances of the vast majority of our sample, the impact of deviating from our chosen mean of 550 pc is minimal. Specifically, at the maximum scale height of 800 pc we would need to produce 11% more stars than at a scale height of 550 pc. Similarly, at 350 pc we would produce 17% more stars than at 550 pc, reducing our SFR. These values are well within our uncertainties seen in Figure 3.7.

A thick disk composed of a range of scale heights has been suggested by recent

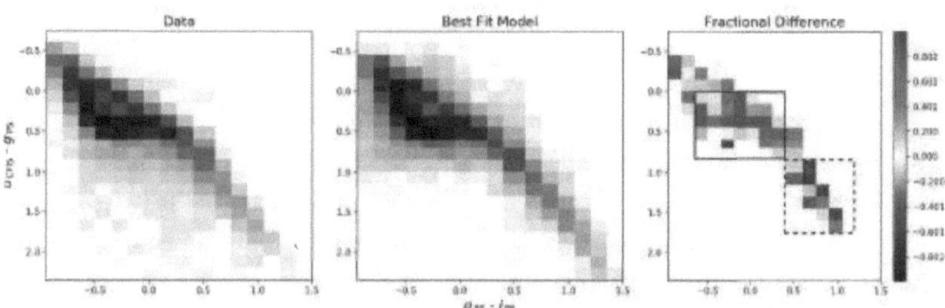

Figure 3.9 Colour-colour diagram showing the data (left) and best-fit model (centre) binned every 0.15 mags. The right-hand panel shows the difference between the data and the model, colour-coded by either an excess (blue) or deficit (red) within the model. The black box shows the main location where the model contains a deficit of white dwarfs relative to the data, and these objects have a mean formation age of (3.3 ± 1.8) Gyr. The dashed box represents the location where our model over-predicts the number of white dwarfs, and these objects have a mean formation age of (5.8 ± 1.1) Gyr. This suggests a more bimodal formation history, with a 50% increase in SFR near 3 Gyr and a 30% deficit at 6 Gyr.

results in order to explain the dip in [α/Fe] seen in the APOGEE data (Bovy et al., 2012; Haywood et al., 2013, 2016). Given that we modeled a single scale height, adding further components to our model would result in both scenarios described above. The net result would be a flattening of the thick disk SFH with a minimal decrease to the peak since the total number of white dwarfs produced would need to remain the same.

3.5.3 Effect of Metallicity

In our model, we have chosen to set a constant metallicity for each Galactic component. The metallicity of a given star affects the pre-white dwarf lifetime through the analytical lifetimes from Hurley et al. (2000). Therefore, changing the metallicity of a given star will change its lifetime, and in turn its white dwarf cooling age. Specifically, for a fixed initial mass, the progenitor lifetime will increase with increasing metallicity, which in turn decreases the white dwarf cooling time. Tononi et al. (2019) found that including a dispersion in the metallicity provided a better fit to the 40 pc white dwarf luminosity function. Adding a dispersion to our metallicity would add a dispersion to the progenitor ages, which, through the cooling age, affects the photometry. If the dispersion is symmetric the effect would be equal at all ages and thus not impact our results. We have studied this effect in Figure 3.10, which shows the fractional change in progenitor lifetime for a higher and lower metallicity relative to our chosen values for the thin and thick disk. While not perfectly symmetric, the implementation of a constant dispersion would only systematically increase our ages by a few percent, which is minimal relative to our uncertainties.

3.5.4 Effect of the IMF and IFMR

Ultimately, as with any population synthesis study, our results are contingent on our assumptions. Assuming that the IMF is constant over time, changing our IMF to a Salpeter or Chabrier will affect the overall normalization but not the shape of our SFH. This is because much of the discrepancy between IMFs is present at the low-mass end and the white dwarfs currently present in the Milky Way were formed by stars with initial massed greater than $0.8-0.9$ M$_\odot$, for which the slope of the IMF is well constrained.

Changing the IFMR will impact the ages of the white dwarfs, as the white dwarf cooling age is a strong function of the mass. As the IFMR is generally estimated

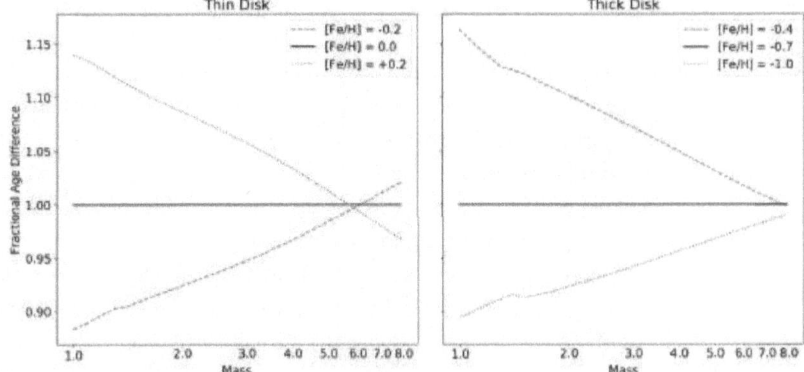

Figure 3.10 Fractional difference between the mean metallicity (black solid line) for the thin disk (left) and thick disk (right) for a star with higher metallicity (dotted blue) and lower metallicity (dashed red) than the mean.

from empirical data, *Gaia* DR2 can be used to calculate the IFMR. Recent work by El-Badry et al. (2018), who used spectroscopically confirmed DAs in wide binaries, has shown promise. Extending this work to all white dwarfs has been challenging given the impact of the atmospheric type on the resulting cooling age, which requires spectroscopic follow-up. The results by El-Badry et al. (2018) and Cummings et al. (2018) also showed that much of the discrepancy between IFMRs arises at high progenitor masses, for which there are very few such stars in our observed sample (see their Figure 3). While we note that there is no consensus on an IFMR, changing the IFMR is likely to have the highest impact on the SFH at old ages where the majority of old white dwarfs are formed from higher mass progenitors.

In order to test this, we ran a simulation with the semi-empirical IFMR derived by Cummings et al. (2018) using the MIST isochrones from Choi et al. (2016) to calculate the progenitor ages. The IFMR presented by Cummings et al. (2018) produces systematically larger white dwarf masses relative to the IFMR from Kalirai et al. (2008), with typical differences being $\sim$0.05 $M_\odot$. At a fixed SFH, or equivalently, fixed white dwarf cooling time, increasing the white dwarf mass will increase the resulting temperature. The end result is that the ages must increase to match the observed colours as these white dwarfs will require more time to cool to equivalent temperatures, represented by the observed colours in our model. The resulting SFH with the Cummings et al. (2018) IFMR can be seen in Figure 3.11. The shape of the two SFHs are roughly consistent, although the resulting thick disk functional age is $\sim$0.3 Gyr older than the result with the Kalirai et al. (2008) IFMR seen in Figure 3.7.

3.5.5 Effect of Atmospheric Composition

In our model, we assumed that white dwarfs have one of two atmospheric compositions: pure-hydrogen ($q_H=M_H/M_\star=10^{-4}$) or helium-rich ($q_H=M_H/M_\star=10^{-10}$). In reality, white dwarfs with helium dominant atmospheres typically contain larger amounts of hydrogen (Koester & Kepler, 2015; Rolland et al., 2018). Genest-Beaulieu & Bergeron (2019) showed that this additional hydrogen impacts the photometric mass determinations of helium-atmosphere white dwarfs, particularly at low temperatures where the hydrogen abundance can increase due to convective mixing. Given that our model obtains masses from the IFMR (and hence both populations have the same mass distribution), this issue will not impact our results. However, a change in the atmospheric composition will affect the rate of cooling of the helium-atmosphere

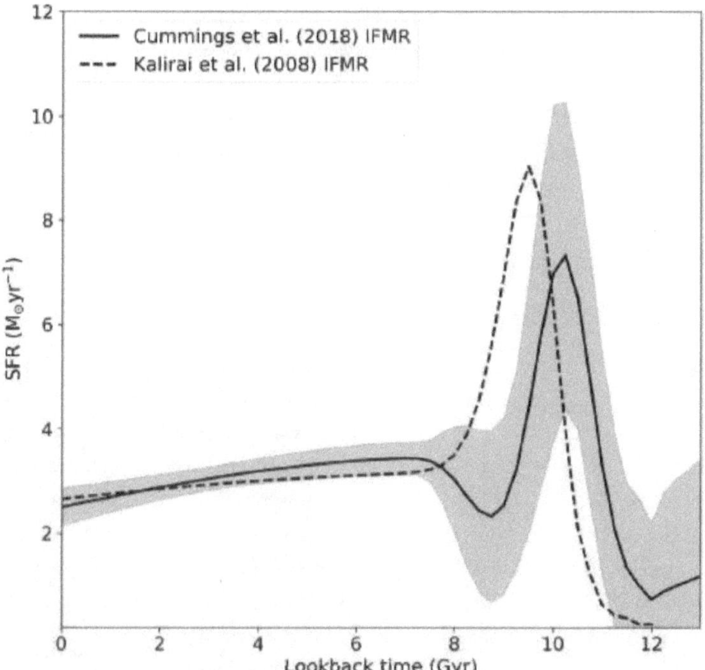

Figure 3.11 A comparison between the resulting star formation history using the IFMR from Kalirai et al. (2008) (dashed line) and Cummings et al. (2018) (solid line). The Cummings et al. (2018) IFMR returns systematically higher masses ($\sim$0.05 M$_\odot$) which results in a 0.3 Gyr increase in the age for the thick disk.

white dwarfs. As discussed in Kilic et al. (2017), pure-helium and pure-hydrogen white dwarfs take the same amount of time to cool to roughly 5,000 K. However, hydrogen-atmosphere white dwarfs begin to cool slower than their helium counterparts below 5,000 K. Specifically, for a 0.6 $M_\odot$ white dwarf with a pure-hydrogen atmosphere, it will take $\sim$1.5 Gyr longer to reach 4,000 K than if it had a pure-helium atmosphere. At 0.8 $M_\odot$, this discrepancy rises to $\sim$2.5 Gyr. Within this temperature range, our resulting model produces very few white dwarfs, and those which pass our selection criteria have pure-hydrogen atmospheres. This is consistent with recent results suggesting that the majority of cool white dwarfs have nearly pure-hydrogen atmospheres (Kowalski & Saumon, 2006; Giammichele et al., 2012; Limoges et al., 2015; Bergeron et al., 2019). The addition of mixed H/He atmospheres may allow a few He-dominant white dwarfs to pass the selection criteria by slowing down their cooling at this epoch, however, this effect would be minimal given the scarcity of objects in this region.

3.5.6 Model Improvements

A few straightforward improvements could be made to our white dwarf population synthesis model in the future. The first would be the inclusion of binaries within the model. Many main sequence+white dwarf binaries would not pass our selection criteria given their colours, which would increase the resulting star formation because the model would need to produce more stars to match the data.

Binary evolution during previous stages of stellar evolution can also affect the evolution of white dwarfs. Binary interactions can strip mass from progenitors forming He-core white dwarfs with masses below those seen in carbon-oxygen white dwarfs. Furthermore, blue straggler formation, or the merging of two white dwarfs, can also lead to an overabundance of higher-mass white dwarfs (Liebert et al., 2005; Parada et al., 2016b; Toonen et al., 2017). This scenario will not only affect the progenitor age, but also the cooling age since higher mass white dwarfs cool more quickly than their lower-mass counterparts. This will inevitably affect the measured star formation history, however, further work will be needed to determine the exact contribution from this scenario.

The presence of an overabundance of white dwarfs with masses between 0.7 - 0.9 $M_\odot$ can also be explained by the onset of crystallization in the interior of the white dwarf Tremblay et al. (2019); Kilic et al. (2020). This effect releases latent heat in the core and causes the cooling to slow down. The result is that white dwarfs will pile up at the temperatures where crystallization occurs. Other effects, such as the sedimentation of ^{22}Ne may also cause a significant delay in the cooling of white dwarfs (Cheng et al., 2019). These effects are not accounted for in the model cooling tracks presented in Figure 1.3, and have been shown to affect the cooling ages of cool white dwarfs (Hollands et al., 2018b; Bergeron et al., 2019; Tremblay et al., 2019; Kilic et al., 2020). Updated models which take these effects into account will be needed to study their effects on the resulting star formation history of the Milky Way.

Extremely low mass (ELM) white dwarfs (M < $0.3\,M_\odot$) may also be present within our dataset. These white dwarfs cannot be formed through normal evolutionary channels and are thought to form as a result of extreme mass-loss before the horizontal branch, leaving an exposed core composed primarily of helium (see, e.g, Kilic et al., 2011; Sun & Arras, 2018). Recent work by Pelisoli & Vos (2019) using *Gaia* DR2 suggests a space density of $275\,\mathrm{kpc}^{-3}$, which would result in only a handful of objects within our sample.

3.5.7 Better Data: A Look Ahead

Finally, we reiterate the need for deeper proper motion surveys in order to constrain the properties of the Galactic halo population. Future surveys like the Large Synoptic Survey Telescope, *WFIRST*, *Euclid* (Laureijs et al., 2011), and *CASTOR* (Côte et al., 2012) will usher in a new era in deep wide-field surveys that will uncover thousands of halo white dwarf candidates.

In order to emphasize this point, we have run a simulation with the mean parameters acquired as part of our fit to the depth of CFIS-u ($u = 24.2$). The resulting colour-colour diagram and RPMD can be seen in Figure 3.12, where the right-hand panel shows the simulation with the CFIS magnitude limit and the left-hand panel is the same as Figure 3.5. The resulting simulation with the CFIS magnitude limit returns half an order of magnitude more white dwarfs, and the change is particularly noticeable in the thick disk and halo populations. Specifically, we expect that our sample of halo objects would increase by a factor of 5−10, and our thick disk sample by a factor of three at the depth of the CFIS observations. Given their importance in understanding Galactic evolution, in particular the age of the halo via the white dwarf luminosity function, upcoming surveys such as LSST and *WFIRST* will provide an unprecedented look into the evolution of the Galactic inner halo via its white dwarf population. Chapter 5 presents an in-depth look into these future surveys and their potential impacts on white dwarfs as tracers of Galactic evolution.

3.6 Summary

In this chapter, we have presented a newly developed white dwarf population synthesis code described in Chapter 2 and used it to determine the star formation history of the Galactic thin disk, thick disk, and stellar halo using new exquisite u-band data from CFIS. Our population synthesis model takes Milky Way geometry and extinction, as well as survey parameters such as completeness, into account in order to return a mock catalogue of white dwarfs in the specified photometric bands. We use data from the Canada-France Imaging Survey (CFIS), Pan-STARRS 1 (PS1) and *Gaia* DR2 to compare to our resulting simulations with a given star formation history, which we have parameterized using skewed Gaussian functions. This sample was shown to be local to the solar neighbourhood, with a median distance of 388 pc. Our main results are the following:

- The star formation history of the Milky Way disk begins (11.3 ± 0.5) Gyr ago with the formation of the thick disk. The thick disk formed stars for 3 Gyr and reached a maximum of $(8.8 \pm 1.4)\,\mathrm{M_\odot yr^{-1}}$ at (9.8 ± 0.3) Gyr.

- The thick disk peak was followed by a decline in the star formation for 1 Gyr before the thin disk began forming stars at a roughly constant rate of $(3.5 \pm 0.3)\,\mathrm{M_\odot yr^{-1}}$ for the past 8 Gyr. The maximum star formation rate of the thin disk occurred at (8.4 ± 0.3) Gyr.

- Although the star formation shape parameters for the stellar halo are relatively poorly constrained, we find a mean age for the inner halo of $10.9^{+1.3}_{-1.1}$ Gyr.

- Studying the residuals reveals variations from a unimodal function, with an increase by as much as 50% at (3.3 ± 1.8) Gyr and a decrease of 30% at (5.8 ± 1.1) Gyr.

- The resulting mass of the thin disk, thick disk, and stellar halo was found to be $(3.4 \pm 0.6) \times 10^{10}\,\mathrm{M_\odot}$.

- The white dwarf space densities were found to be $(4.8 \pm 0.4) \times 10^{-3}\,\mathrm{pc^{-3}}$ for the thin disk, $(1.0 \pm 0.2) \times 10^{-3}\,\mathrm{pc^{-3}}$ for the thick disk, and $(6.3 \pm 2.4) \times$

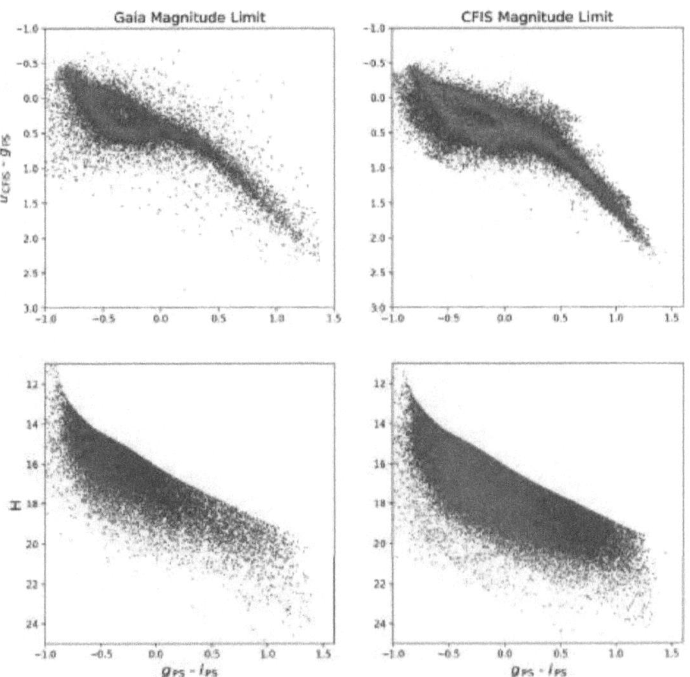

Figure 3.12 Comparing the resulting white dwarf populations based on the *Gaia* magnitude limit (left columns; $G = 20.7$) and the CFIS magnitude limit (right columns; $u = 24.2$). The fainter magnitude limit results in a $5-10\times$ increase in halo white dwarfs relative to the CFIS-PS1-*Gaia* sample.

10^{-6} pc^{-3} for the halo. The fractional contributions of the thin disk, thick disk, and halo to the local white dwarf population were found to be $(83 \pm 5)\,\%$, $(17 \pm 3)\,\%$, and $(0.11 \pm 0.05)\,\%$ respectively.

- We find a He-fraction of $(21 \pm 3)\,\%$. This is consistent with previous studies of the solar neighbourhood.

Given the *Gaia*-imposed magnitude limit of our survey, our dataset is dominated by disk stars in the solar neighbourhood. Future deep proper motion surveys will be needed to increase the sample of halo white dwarfs in order to properly constrain its formation parameters. Upcoming surveys, like Pan-STARRS1 DR2, the Large Synoptic Survey Telescope, as well as the Roman Space Telescope, Euclid, and CASTOR if they incorporate proper motions will open a new window into the formation of our Milky Way at its earliest times.

Chapter 4

The Mass and Age Distribution of High-Velocity White Dwarfs

4.1 Introduction

The Galactic halo contains the most ancient stars in the Milky Way, and as such provides a means to study the early formation and evolution of our Galaxy. Many of these studies benefit from an accurate measurement of the age of a star, however, age cannot be measured directly, but instead relies on theoretical or semi-empirical models (Soderblom, 2010). As described in Chapter 1, the evolution of an isolated white dwarf is determined by its mass, temperature, and atmospheric composition, and as such, the determination of these properties can be used to infer a white dwarf cooling age. These properties can be measured either spectroscopically or photometrically provided an accurate parallax is available (Bergeron et al., 2005; Kalirai, 2012; Kilic et al., 2019).

Various methods using white dwarfs have been used to determine the ages of stellar populations of all ages, from the oldest globular clusters in the halo to younger open clusters in the disk, as well as field stars belonging to different components of the Milky Way (see, e.g, Winget et al., 1987; Hansen et al., 2007; Bedin et al., 2009; Tremblay et al., 2014; Kilic et al., 2019). Typically these studies rely on the white dwarf luminosity function as white dwarfs will accumulate at cool temperatures given the finite age of the population (Harris et al., 2006; Rowell & Hambly, 2011; Kilic et al., 2017). These methods rely on the observations of the oldest white dwarfs, which are cooler and fainter than their younger counterparts which makes them difficult to detect in optical surveys.

The Galactic halo, however, is also continuously producing young, hot, white dwarfs, but with lower masses than their disk counterparts. This is because they evolve from older, lower mass, progenitors (Kalirai, 2012). Measuring the age of these white dwarfs has presented some unique challenges. Since the total age of the star is the sum of the white dwarf cooling age and the pre-white dwarf lifetime of its progenitor, both must be derived. The white dwarf cooling age can be determined from the mass, temperature, and atmospheric composition. The difficulty lies in measuring the progenitor lifetime, which dominates the total age of these young white dwarfs, as it relies on the ability to relate the observed white dwarf mass to the progenitor mass via the IFMR.

The IFMR can be determined theoretically by integrating mass loss rates over the lifetime of a star using stellar evolution models (see, e.g, Weidemann, 2000; Marigo & Girardi, 2007; Choi et al., 2016), however, previous studies have frequently relied on observational methods. The IFMR can be determined using star clusters as the homogeneous age of their stars means that the main-sequence turn-off mass can be directly related to the mass of the youngest white dwarfs (see, e.g, Kalirai et al., 2008; Cummings et al., 2018), but wide double degenerate binaries in the field Andrews et al. (2015) or white dwarfs with accurate parallaxes El-Badry et al. (2018) have also been used.

There exist discrepancies in the various IFMRs as a result of the varying methods, however, much of the disagreement occurs at either high or low initial mass — the most poorly sampled regions of parameter space. In the high mass regime, there exist very few young open clusters with significant white dwarf populations. Furthermore, these high-mass progenitors are less numerous, have short lifetimes, and produce white dwarfs that are intrinsically fainter.

At low mass (important for the young Galactic halo white dwarfs) there exist very few nearby globular clusters for which spectroscopic white dwarf masses can currently be obtained. Currently, the low-mass IFMR is anchored by spectroscopic measurements of just a single globular cluster, Messier 4 (M4), performed by Kalirai et al. (2008), which revealed white dwarf masses between 0.52 and 0.58 $M_\odot$. Kalirai

(2012) used the measured masses of four halo white dwarfs, combined with this IFMR to infer an inner halo age of $(11.7 \pm 0.7)\,\mathrm{Gyr}$.

Given that the age of the halo remains uncertain given the small sample size of the Kalirai (2012), it is not surprising that the SFH of the halo also remains uncertain. In Chapter 3, we concluded that the photometric sample of halo objects in our CFIS dataset was insufficient, particularly relative to the disk, to measure an accurate SFH. One solution is to acquire spectroscopic temperatures and masses in order to directly measure the age distribution of a sample of halo white dwarfs. In this chapter, we build on the sample of Kalirai (2012) by acquiring Gemini-GMOS spectroscopic observations of 18 halo white dwarfs in order to determine their age and mass distributions.

In Section 4.2, we detail our selection method as well as all photometric and spectroscopic observations. We present the resulting masses and ages in Section 4.3 and discuss their implications in Section 4.4 before summarizing and looking towards the future in Section 4.5.

4.2   Data

4.2.1   Photometric Data and Selection

This chapter uses the CFIS-PS1-*Gaia* white dwarf catalogue presented in Chapter 3. In this work, we target the white dwarfs with the most extreme tangential velocities — those most unambiguously associated with the halo. Our targets were selected based on their location in the RPMD (See Figure 4.2). Given that white dwarfs are intrinsically fainter than their main-sequence counterparts at equal temperatures they will experience a larger proper motion and end up with larger reduced proper motions. This method can cleanly separate white dwarfs from other point-sources and previous studies, for example, have found a contamination rate ranging from $1 - 5\,\%$ from halo sub-dwarfs (see, e.g, Kilic et al., 2006; Dame et al., 2016).

In Figure 4.2 we have plotted model cooling tracks with tangential velocities of $20\,\mathrm{kms}^{-1}$, $40\,\mathrm{kms}^{-1}$, and $200\,\mathrm{kms}^{-1}$, which are typically associated with the mean tangential velocity of the thin disk, thick disk, and stellar halo respectively. The synthetic magnitudes have been calculated in the CFIS-u and PS1 *grizy* bands for H- and He-atmosphere models as in Chapter 3.

An object is considered to be part of the Galactic halo if it lies below the model halo track (i.e $v_t > 200\,\mathrm{kms}^{-1}$). Applying this selection to our dataset results in 90

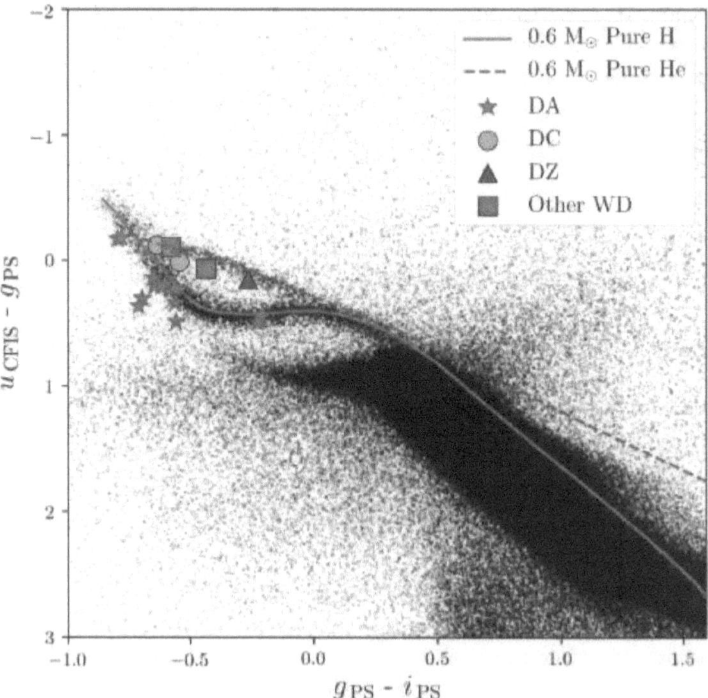

Figure 4.1 Colour-colour plot of all CFIS sources ($u < 21$) with our 2019A halo sample highlighted based on their spectral type. Red and blue lines represent model cooling tracks for $0.6\,\mathrm{M_\odot}$ white dwarfs with hydrogen and helium atmospheres respectively.

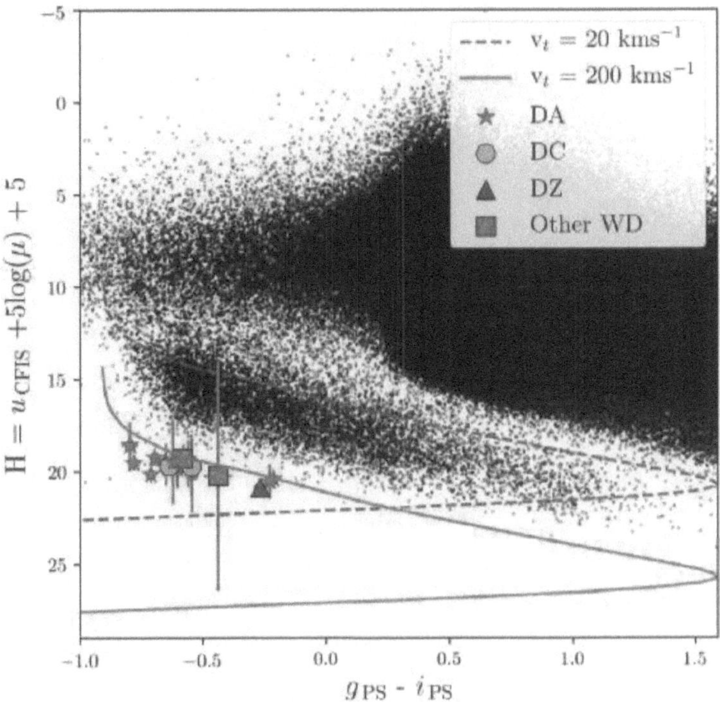

Figure 4.2 Reduced Proper Motion Diagram (RPMD) for our CFIS-PS1-*Gaia* objects ($u < 21$). Also included are model tracks for the thin disk ($v_t = 20\,\mathrm{kms}^{-1}$), and stellar halo populations ($v_t = 200\,\mathrm{kms}^{-1}$). Our halo white dwarf candidates were selected if their reduced proper motion was greater then the $200\,\mathrm{kms}^{-1}$ curve.

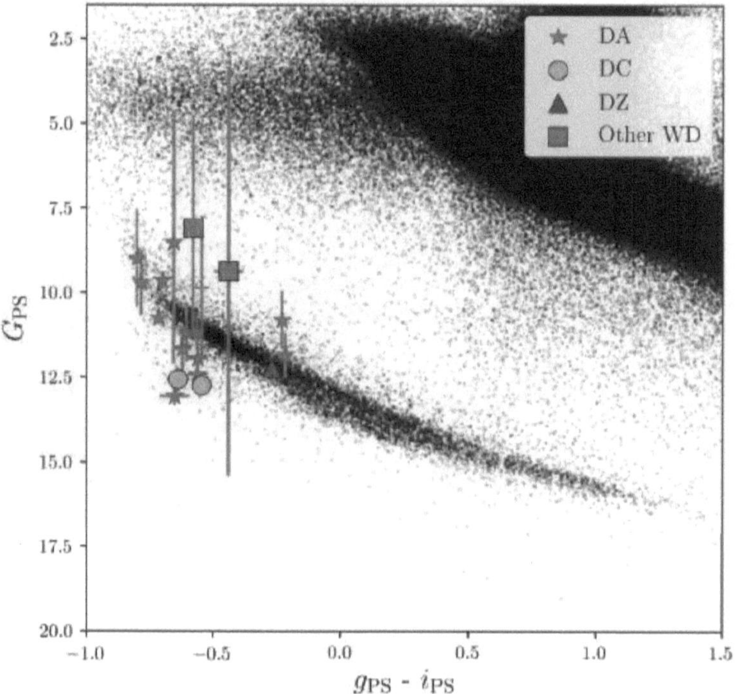

Figure 4.3 Colour-magnitude diagram for the observed white dwarfs with symbols denoting their spectral type. Also plotted are objects from *Gaia* DR2 to highlight the separation between the white dwarf cooling sequence and other point-sources. Distances are calculated using the *Gaia* DR2 parallaxes.

halo white dwarf candidates, however, we target 18 of the brightest objects which were observable during the 2019A semester. We then checked their location in the colour-magnitude diagram, seen in Figure 4.3 to makes sure they lay along the white dwarf cooling track. As Figure 4.3 shows, the objects all lie within the white dwarf region, well separated from the concentration of main-sequence stars seen in the upper right-hand area. Furthermore, we check whether these objects have velocities consistent with the disk of the Milky Way. This check can be seen in Figure 4.4, which shows a Toomre diagram of the objects, with 5σ velocity ellipsoids for the thin and thick disk shown in green and blue respectively. While a few objects could be part of the thick disk at the 3σ level, radial velocity measurements acquired from the resulting spectra show that only two objects lie within the 5σ ellipsoid of the thick disk. This will be discussed in more detail in Section 4.4.1.

4.2.2 Spectroscopic Follow-up

Long-slit spectroscopic follow-up observations were performed using the Gemini Multi-Object Spectrographs (GMOS) at the twin 8-m Gemini Observatories on Mauna Kea and Cerro Pachón. To achieve the desired spectral range, the B600 grating with a central wavelength of 5200 Å was used. This allowed for the simultaneous acquisition of H_α through to H_8, which is ideal for mass and temperature determinations. A focal-plane unit (slit width) of 0.″75 was selected in order to maximize the SNR while maintaining a suitable resolution. The data were binned 2x2, resulting in a resolution of R~1200.

Using the recommendation from Kepler et al. (2006) we set our exposure times so as to acquire spectra with a signal-to-noise ratio (SNR) of 15 per resolution element at the location of the higher-order Balmer lines (H_8) to obtain accurate masses. The resulting integration times were split into sub-exposures in order to minimize the impact of cosmic rays. Given that we were awarded Band 2 time, however, the variable observing conditions resulted in a variety of observed SNR values (see Table 4.2).

A raw spectral image can be seen in Figure 4.5. These resulting images were then processed using the Gemini PyRAF package to remove sources of noise. The main steps in this process are as follows:

- The process begins by removing the constant voltage applied to the detector (a charged-coupled device; CCD) by median combining an odd number of zero-

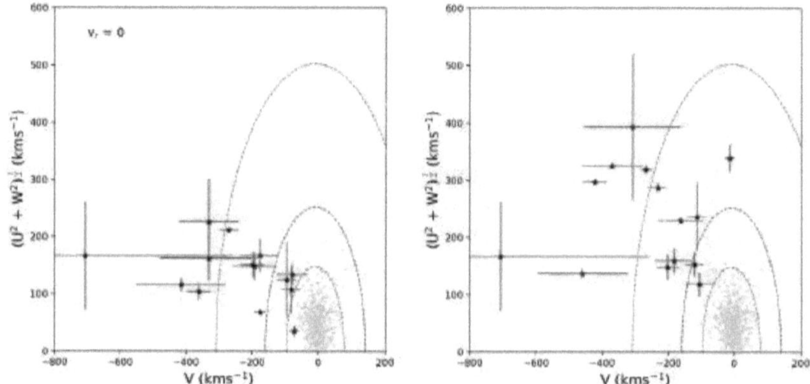

Figure 4.4 *Left:* Toomre diagram for the observed DA white dwarfs assuming zero contribution from the radial velocity *Right:* Using the measured radial velocity from our Gemini spectra.

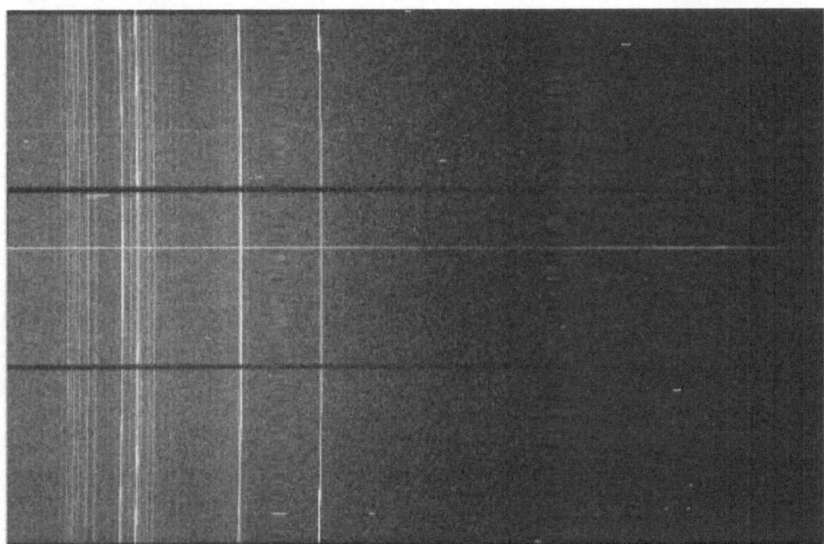

Figure 4.5 Raw spectrum of J0823+3111. The x-axis is the wavelength direction and the y-axis is the spatial direction. The spectrum is displayed in the middle of the y-axis and spans the entire range of the x-axis. Bright spots represent cosmic ray strikes on the detector. Dark horizontal lines and vertical blurred regions are the locations of physical gaps between the detectors. Bright vertical lines indicate strong emission from the Earth's atmosphere.

second exposures. These images are called bias frames and are typically taken at the beginning of an observing night. This master bias is then subtracted from the science images.

- Next, pixel-to-pixel variations are corrected for by using flat field exposures. These images are obtained by equally illuminating each pixel of the CCD. These images were obtained using a lamp, although the sky can also be used during twilight. Since each pixel should be illuminated equally, dividing the science image by a normalized flat will remove any pixel-to-pixel variations.

- The next two steps involve calibrating the x and z axes, shown in Figure 4.5, which represent the wavelength and counts. The wavelength calibration is done by observing a Copper-Argon (CuAr) lamp which contains a number of emission lines at well-determined wavelengths spread over the region of interest. This allows the pixel numbers to be converted to wavelength. The counts are then converted to units of flux by comparing them to a spectral standard. In this case, the spectral standard is an A-type star, Feige 66, which had been observed using an identical instrument setup and shared many similar absorption features with the observed white dwarfs.

- The sky emission is removed by selecting a region of the image away from the spectrum, taking a median along the spatial axis, and subtracting this result from the spectral region.

- The 1-dimensional spectrum is extracted by tracing the peak of the spatial pixels along the wavelength direction. The pixels typically follow a Gaussian profile along the spatial direction, and pixels falling within a defined window are median combined to produce a flux at a given wavelength. An example is shown in Figure 4.6 that shows a vertical slice along the image presented in Figure 4.5. The pixels containing flux from the star are located within the spike at pixel $\sim$1100.

- All of the objects observed in this Chapter were split into multiple sub-exposures in order to minimize the effect of cosmic ray strikes. Thus, the final step is to median combine the resulting 1-D spectra into a single spectrum.

Our sample contains 13 DA, 1 DZ, 2 DC, and 2 peculiar spectra which will be discussed in the following subsection. The DA and DZ spectra can be seen in Figure

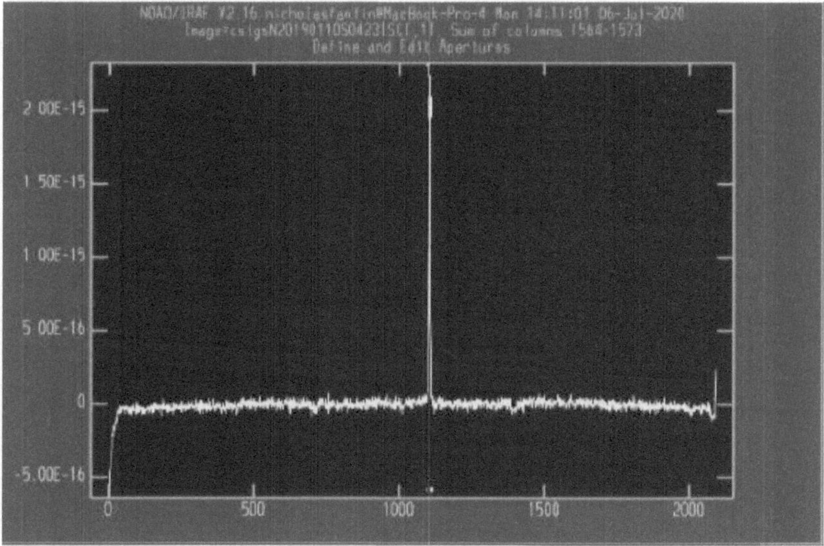

Figure 4.6 Vertical slice along the image presented in Figure 4.5. The x-axis is pixels along this axis and the y-axis is the flux per pixel. The pixels containing flux from the object are represented by a roughly Gaussian distribution seen at pixel ∼1100.

4.7, while the remaining DC and other white dwarf types are seen in Figure 4.8. Also shown in Figure 4.7 are the gaps in the detector (grey shaded region) and the rest frame Balmer series from H_α to H_8 (dashed lines).

4.2.3 Peculiar Spectra

Two of the spectra we obtained contained features inconsistent with being either a DA, DC, or DZ. These objects can be seen in Figure 4.1 as green squares. The observational parameters for these objects (J1637+3631 and J1400+2036) can be found in Table 4.1. Furthermore, their spectra can be seen in Figure 4.8, with a number of common white dwarf atmospheric lines shown for clarity.

One of our objects, J1637+3631 was studied in detail in Raddi et al. (2019). This object shows evidence of oxygen, magnesium, silicon, and calcium absorption, which are peculiar elements to find within a white dwarf atmosphere. The authors suggest that this star is a white dwarf, likely a subtype DS (oxygen features present. see, Williams et al., 2019) given the abundance of oxygen in its atmosphere, the second of its kind after J1240+6170 (Kepler et al., 2016b). These stars are thought to be partly burnt white dwarfs which survived a supernova from a binary companion. Similar spectral features are also seen in J1400+2036, suggesting that it could also belong to this class of remnants. The objects both show large tangential velocities (300-500 kms^{-1}), which would allow them to pass our selection criteria. Given the relative scarcity of these objects and the difficulty in modelling their atmospheric composition, we do not attempt to determine the mass or age of these objects. Therefore, we continue our analysis using the DA, DC, and DZ objects unless otherwise noted. It should be noted, however, that based on our sample these objects may be more common amongst the high-velocity sample than previously expected. Future large-scale spectroscopic studies, such as WEAVE (Dalton et al., 2012), will be needed to further characterize the prevalence of such objects.

4.3 Determination of the Mass, Temperature, Age, and Uncertainties

4.3.1 Temperature and Mass

DA

We obtain temperatures ($T_{\rm eff}$) and surface gravities ($\log g$) by fitting the observed Balmer lines with model atmospheres (Bergeron et al., 1992). The resulting temper-

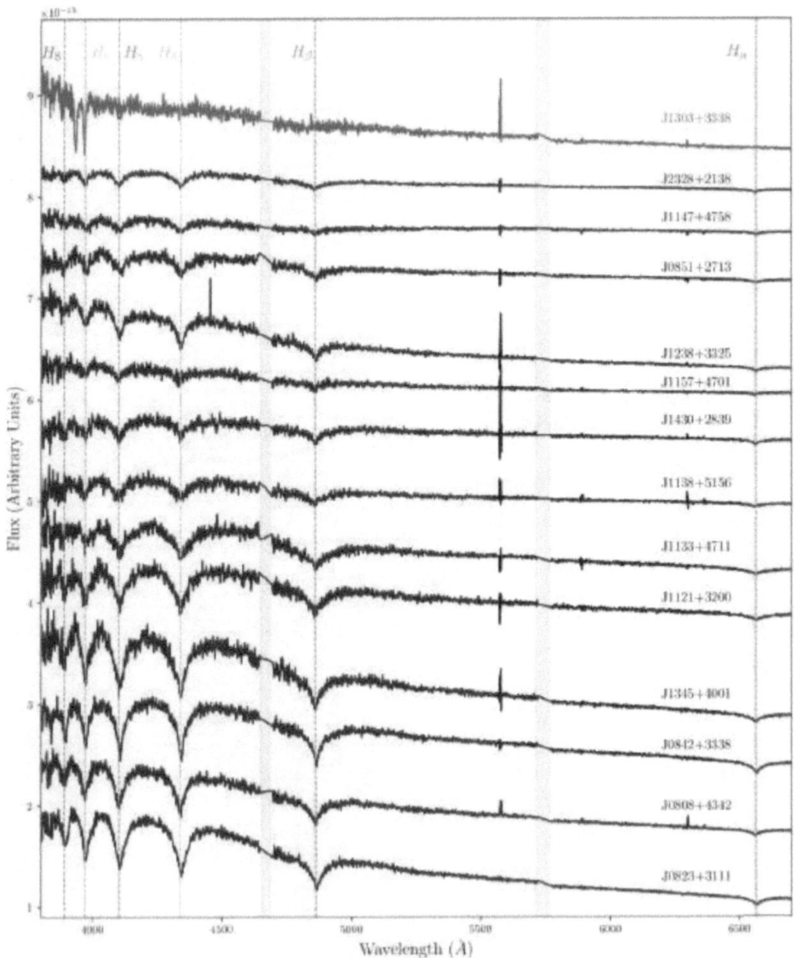

Figure 4.7 Spectra of 14 high-velocity white dwarfs obtained using GMOS on Gemini-North for which a spectroscopic mass can be determined. The top spectrum, J1303-3338, is a DZ with prominent Ca H &K absorption, while the remaining are classified as DA. The Balmer series is marked for clarity by the dashed lines. The chip gaps on the GMOS detector are located within the shaded grey regions.

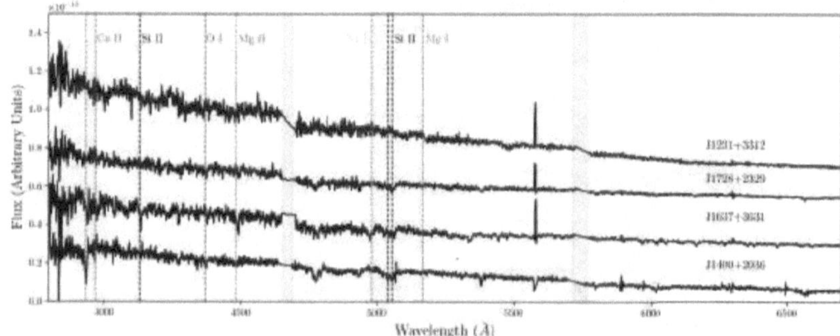

Figure 4.8 GMOS spectra of the four non-DA or DZ white dwarfs. J1637+3631, a remnant of a peculiar thermonuclear reaction presented in detail by Raddi et al. (2019), is shown as the second from the bottom. A number of common absorption features are highlighted to guide the eye. The chip gaps on the GMOS detectors are located within the shaded grey regions as in Figure 4.7.

Table 4.1 Photometric properties of halo white dwarfs observed with GMOS

ID	RA (J2000) HH:MM:SS	Dec (J2000) $\circ$:$\prime$:$\prime\prime$	u_{CFIS} AB mag	g_{PS} AB mag	i_{PS} AB mag	μ_α $\prime\prime$/yr	μ_δ $\prime\prime$/yr	parallax mas
2	08:08:48.96	43:42:14.40	20.04 ± 0.01	20.04 ± 0.02	20.30 ± 0.02	-4.5 ± 1.5	-77.1 ± 0.8	0.49 ± 0.82
3	08:23:59.15	31:11:52.10	18.31 ± 0.00	18.11 ± 0.00	18.81 ± 0.02	-15.4 ± 0.3	-165.5 ± 0.2	2.02 ± 0.23
4	08:42:49.37	33:38:43.50	20.22 ± 0.01	19.81 ± 0.01	20.04 ± 0.01	-67.4 ± 0.9	-121.3 ± 0.5	1.58 ± 0.63
5	16:37:12.21	36:31:55.90	19.99 ± 0.01	20.14 ± 0.02	20.71 ± 0.02	-18.5 ± 1.2	64.3 ± 1.2	0.39 ± 0.60
6	08:51:15.42	27:13:54.30	19.69 ± 0.01	19.47 ± 0.01	20.04 ± 0.01	-21.1 ± 0.8	-90.0 ± 1.6	2.17 ± 0.52
10	11:21:31.34	32:00:12.30	20.27 ± 0.01	19.86 ± 0.01	20.42 ± 0.04	36.6 ± 1.4	-77.6 ± 2.1	2.61 ± 1.21
12	11:33:27.50	47:11:36.50	19.87 ± 0.01	19.73 ± 0.02	20.38 ± 0.05	-73.4 ± 0.7	-59.6 ± 0.9	4.62 ± 0.57
13	11:38:59.14	51:56:47.70	20.34 ± 0.01	20.04 ± 0.02	20.60 ± 0.03	27.4 ± 0.8	-79.3 ± 0.7	2.98 ± 0.62
14	11:47:55.58	47:58:13.80	20.13 ± 0.01	20.02 ± 0.01	20.62 ± 0.02	-82.2 ± 0.5	-43.8 ± 0.8	1.70 ± 0.70
15	11:57:51.05	47:01:23.10	19.32 ± 0.01	19.57 ± 0.01	20.35 ± 0.02	-32.5 ± 0.5	-92.9 ± 0.5	1.06 ± 0.47
16	12:31:25.29	33:12:13.10	19.51 ± 0.01	19.51 ± 0.02	20.06 ± 0.02	-105.4 ± 0.8	33.3 ± 0.5	4.41 ± 0.46
17	12:38:05.70	35:25:23.70	19.15 ± 0.01	19.36 ± 0.01	20.16 ± 0.03	-57.8 ± 0.7	-36.0 ± 0.5	0.84 ± 0.56
21	13:03:34.31	33:38:43.30	20.09 ± 0.01	20.04 ± 0.03	20.30 ± 0.02	-3.9 ± 1.0	-145.2 ± 0.9	2.74 ± 0.52
26	13:45:51.89	40:01:00.90	18.73 ± 0.01	18.63 ± 0.01	19.34 ± 0.02	-192.4 ± 0.2	-52.8 ± 0.3	2.65 ± 0.22
27	17:28:02.89	23:29:12.20	19.90 ± 0.01	20.02 ± 0.03	20.66 ± 0.03	6.8 ± 0.6	-85.6 ± 0.8	3.20 ± 0.51
28	14:30:32.55	28:39:11.10	20.44 ± 0.01	20.03 ± 0.03	20.25 ± 0.02	-10.6 ± 1.3	-117.0 ± 1.4	2.28 ± 0.66
2018B	23:28:55.81	21:38:26.4	20.05 ± 0.01	19.92 ± 0.01	20.54 ± 0.03	76.2 ± 1.1	-27.8 ± 0.8	2.22 ± 0.69
30	14:00:11.93	20:36:54.40	20.45 ± 0.01	20.56 ± 0.03	21.00 ± 0.05	-85.8 ± 2.9	1.3 ± 3.9	0.58 ± 1.61

Note. — WD 30 was observed using Gemini South

Table 4.2 Properties from Spectral and Photometric Fitting

			Spectroscopy			Photometry[2]		
ID	SNR[1]	Type	$T_{\rm eff}$(K)	Mass ($M_\odot$)	t_{cool} (Myr)	$T_{\rm eff}$(K)	Mass ($M_\odot$)	t_{cool} (Myr)
J0808+4342	15.6	DA	22207 ± 543	0.531 ± 0.036	28	21209 ± 1122	$0.194^{+0.41}_{-0.283}$	74
J0823+3111	20.2	DA	19764 ± 366	0.485 ± 0.026	44	18562 ± 754	$0.305^{+0.040}_{-0.034}$	42
J0842+3338	19.3	DA	10779 ± 182	0.589 ± 0.044	481	10569 ± 445	$0.238^{+0.159}_{-0.158}$	212
J1637+3631	10.0	DS[3]				21209 ± 1122	$0.194^{+0.41}_{-0.283}$	18
J0851+2713	10.5	DA	13220 ± 684	0.757 ± 0.088	410	16605 ± 683	$0.648^{+0.186}_{-0.209}$	155
J1121+3200	13.6	DA	13835 ± 868	0.734 ± 0.077	341	13489 ± 925	$0.839^{+0.278}_{-0.488}$	475
J1133+4711	13.0	DA	16714 ± 1820	1.228 ± 0.040	956	16545 ± 1061	$1.25^{+0.040}_{-0.055}$	1023
J1138+5156	7.8	DA	16588 ± 928	0.854 ± 0.098	276	14538 ± 918	$1.055^{+0.112}_{-0.174}$	706
J1147+4758	9.5	DA	15217 ± 1024	0.684 ± 0.100	228	17689 ± 808	$0.720^{+0.281}_{-0.368}$	157
J1157+4701	9.0	DA	30584 ± 898	0.581 ± 0.083	9	26842 ± 1437	$0.511^{+0.278}_{-0.207}$	13
J1231+3312	15.1	DC				18022 ± 819	$1.214^{+0.040}_{-0.053}$	756
J1238+3325	15.0	DA	28129 ± 774	0.628 ± 0.065	12	27663 ± 1488	$0.384^{+0.302}_{-0.214}$	13
J1303+3338	13.5	DZ	10635 ± 363	$0.664^{+0.197}_{-0.162}$	649	12517 ± 640	$0.829^{+0.145}_{-0.188}$	
J1345+4001	13.6	DA	17819 ± 436	0.569 ± 0.043	107	18401 ± 787	$0.609^{+0.074}_{-0.072}$	93
J1728-2329	9.5	DC				23573 ± 1284	$1.255^{+0.046}_{-0.0.69}$	412
J1430+2839	8.9	DA	11050 ± 327	0.657 ± 0.095	525	10223 ± 381	$0.477^{+0.229}_{-0.206}$	432
J1400+2036	20.1	Unknown				19890 ± 1112	$0.277^{+0.846}_{-0.169}$	1011
J2328+2138	17.0	DA	18918 ± 514	0.571 ± 0.046	70	18284 ± 833	$0.891^{+0.196}_{-0.315}$	229

Note. — [1] SNR reported as the average SNR per pixel between 4400 and 4600 $\mathring{A}$

Note. — [2] Photometric values assume [He/H] = 0

Note. — [3] See Raddi et al. (2019) for further details on classification

ature and surface gravity values are corrected for 3D effects based on the results of
Tremblay et al. (2013) before being converted to a mass using the white dwarf mass-
radius relationship. This technique has been used in a number of previous studies
to measure the masses of DA white dwarfs (see, e.g, Tremblay et al., 2011; Genest-
Beaulieu & Bergeron, 2019). An example of the fitting technique for the DA sample
can be seen in Figure 4.9.

The reported errors on the white dwarf parameters take both internal and external
errors associated with the fitting technique. The internal errors are based on how
well the model fits the data, with this error approaching zero as the SNR approaches
infinity. The external errors are estimated to be 1.2% in T_{eff} and 0.038 dex in log g
as determined by Liebert et al. (2005). The total error is then the sum in quadrature
of the internal and external errors.

We apply this technique to our sample of 13 DA white dwarfs and present the
results in Table 4.2.

DZ

Analysis of the DZ (J1303+3338) was performed in the same manner as described in
Coutu et al. (2019) based on updated methods from Dufour et al. (2005, 2007). This
method uses a combination of the photometric method in combination with a metallicity determined using the spectrum. The resulting fit also returns the temperature
and mass, which we also present in Table 4.2.

4.3.2 Photometric Technique

Another technique that can be used to determine the surface temperature and mass
of a white dwarf relies on the distance and spectral energy distribution. This technique, called the photometric technique, uses the theoretical magnitudes from model
atmospheres and compares them to broadband photometric observations (Bergeron
et al., 1997). For this method we use the CFIS u in combination with the PS1 *grizy*
bands. The u-band in particular is important for determining the temperatures of
hot white dwarfs (Genest-Beaulieu & Bergeron, 2014) as they have a considerable
amount of ultraviolet flux (Bergeron et al., 2019).

The best fit returns the temperature and the solid angle, $\pi(R/D)^2$. Therefore,
the parallaxes from *Gaia* DR2 can be used to determine the distance, allowing one
to solve for the radius. The mass is then determined using the mass-radius relation

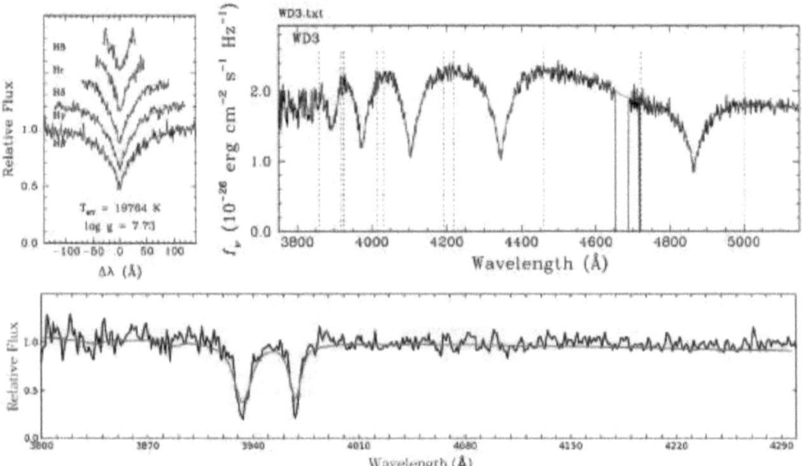

Figure 4.9 *Top:* Fit to the spectrum obtained from GMOS for J0823-3111. The left-hand panel shows the normalized Balmer lines, from H$_\beta$ to H$_8$, while the right-hand panel shows the model fit (red) over-plotted on the observed spectrum (black). *Bottom:* Fit to J1303-3338, a DZ which displayed strong Ca H & K absorption features.

as in the spectroscopic technique. We present these results in the final two columns of Table 4.2.

This procedure can be seen in Figure 4.10, which shows the photometry as error bars and the resulting synthetic photometry as black circles. The left-hand panel shows the results for a hot white dwarf (J1238+3325), while the right-hand panel shows a fit for a cooler white dwarf (J1430+2839). The resulting temperature of both objects agrees with the spectroscopic values, and we note that when the u-band is removed from the fit the photometric temperature is determined to be nearly 10,000 K hotter for J1238+3325.

The resulting photometric temperatures and masses are then compared to their spectroscopic counterparts in Figure 4.11. The temperatures agree rather well between both methods, however, the masses show much larger scatter given the large uncertainties in the $Gaia$ parallax values for some of the objects. The photometric technique does correctly pick out the most massive white dwarf in our sample (J1133+4711), but many of the remaining objects have essentially unconstrained masses due to poor parallax measurements. For this reason, the remainder of the chapter will focus on the resulting masses obtained spectroscopically unless otherwise noted. This leaves our sample at 14 objects, with 13 DA and one DZ, as DC white dwarfs do not contain any spectroscopic features.

4.3.3 Mass Distribution

The resulting spectroscopic and photometric masses are presented in Table 4.2. In the left-hand panel of Figure 4.12 we show the distribution of white dwarf masses by representing each measurement as a Gaussian with a mean equal to the measured mass and the standard deviation equal to the corresponding uncertainty. The distribution peaks at $0.58\,M_\odot$ and contains a few higher and low mass objects, including J1133+4711 with a mass of $1.22 \pm 0.04\,M_\odot$, and J0823+3111 with a mass of $0.49 \pm 0.03\,M_\odot$. Under the canonical assumption that the halo formed $\sim$12 Gyr ago in a short burst of star formation, one would expect these young white dwarfs to have masses of $0.52 - 0.58\,M_\odot$ (Kalirai et al., 2008), and hence such a large spread was not expected.

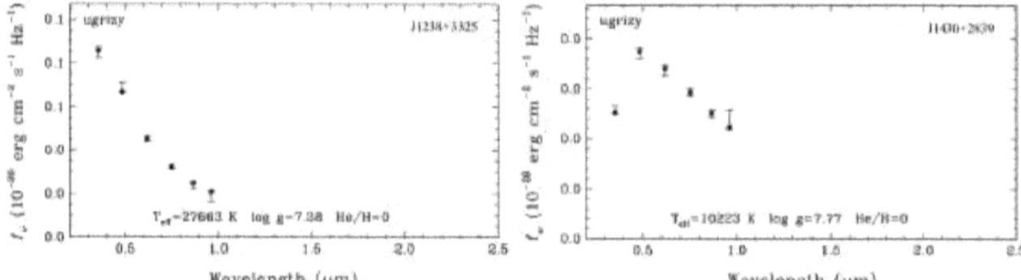

Figure 4.10 An example of the photometric technique used to measure the mass and temperature using the photometry in combination with the *Gaia* parallax. The photometry used includes the CFIS *u* in combination with the PS1 *grizy* bands. The observed magnitudes are represented by the error bars, and the photometry of the best fit model is indicated by the circles. The resulting best fit model parameters are also presented.

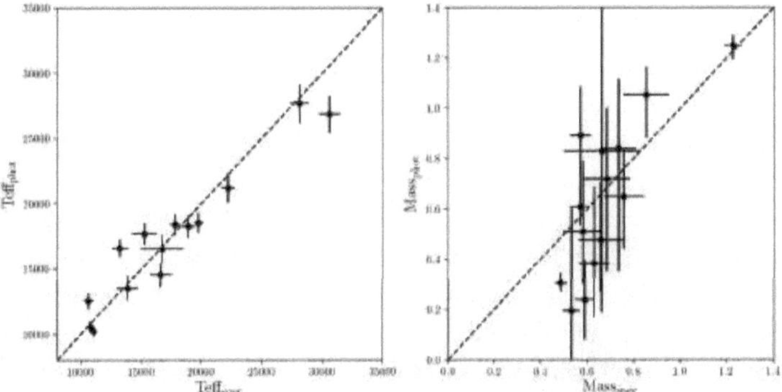

Figure 4.11 Comparing the spectroscopic and photometric temperature (left) and mass (right) determinations for the 14 DA and DZ white dwarfs with spectroscopic masses.

4.3.4 Ages

Ages using Stellar Isochrones

Surprisingly, many of the objects do not have masses consistent with those of white dwarfs in the oldest globular clusters (M $\sim 0.55\,\mathrm{M_\odot}$), with their higher masses suggesting they formed from higher mass progenitors and would not have ages consistent with what is normally believed to be an old halo population.

In order to calculate the total ages we require a progenitor lifetime to add to the calculated white dwarf cooling times presented in Table 4.2. We begin by calculating the progenitor mass using the IFMR. We choose to compare two IFMRs, those from Kalirai et al. (2008) and Cummings et al. (2018). The main difference between the two is that for a given initial mass the Cummings et al. (2018) IFMR returns systematically higher white dwarf masses on the order of $0.05\,\mathrm{M_\odot}$. Equivalently, this means that for a given white dwarf mass the Cummings et al. (2018) IFMR will return lower initial masses, and hence longer pre-white dwarf ages. The resulting initial masses for both IFMRs can be seen in the middle and right-hand panels of Figure 4.12.

With the initial mass calculated, the remaining step is to calculate the progenitor's lifetime. This requires an assumption regarding the metallicity before using a stellar isochrone to determine the progenitor age. For this study, we use two isochrones: the PAdova and tRieste Stellar Evolution Code (PARSEC) (Bressan et al., 2012) and the Mesa Isochrones and Stellar Tracks (MIST) (Choi et al., 2016). We adopt an initial metallicity of [Fe/H] $= -1.5$ dex for our progenitors, which is typical for a halo population (Peng et al., 2013).

With a choice of metallicity and isochrone, we calculate the progenitor age and add it to the white dwarf cooling age to obtain the total age of each star. We then propagate the uncertainties in the white dwarf mass and cooling ages through the IFMR and stellar isochrone to obtain the distribution of total ages and present the sum of these histograms in Figure 4.13. We note that the choice of IFMR has a much larger effect on the total age relative to the choice of isochrone, and as such we will only present results with the MIST isochrones. Each object is plotted as a Gaussian with the mean equal to the age, and the standard deviation equal to the error in the age. We also present the resulting ages using the various combination of isochrones and IFMRs in Table 4.3.

Our results show that the majority of our sample have ages inconsistent with other estimates of the age of the Galactic halo, irrespective of our choice of isochrone or

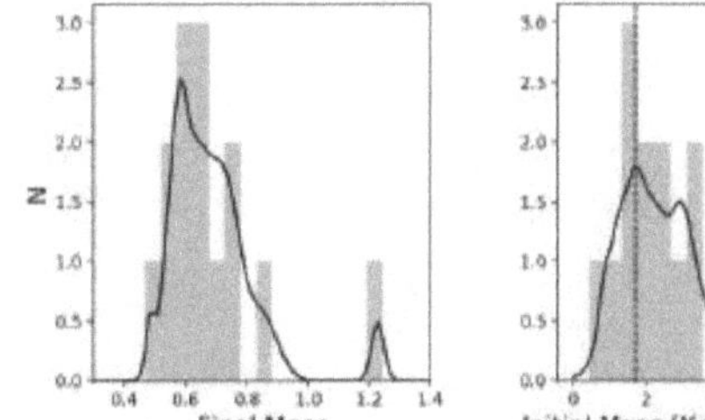
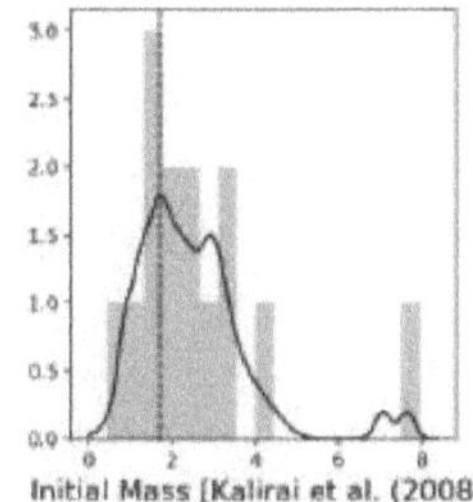
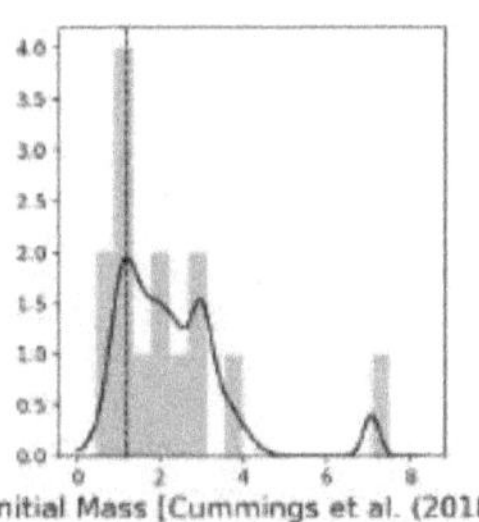

Figure 4.12 *Left:* The distribution of white dwarf masses as displayed in Table 4.2. *Middle:* Initial masses calculated using the Cummings et al. (2018) IFMR for our sample of white dwarfs. The mean initial mass for the sample is indicated by a dashed line. *Right:* Same as the middle panel except the IFMR of Kalirai et al. (2008) was used.

Table 4.3 Ages for Halo White Dwarfs

ID	T_{eff}(K)	Mass (M$_\odot$)	t_{cool} (Myr)	Age (Gyr) MIST + K08	Age (Gyr) PARSEC + K08	Age (Gyr) MIST + C18	Age (Gyr) PARSEC + C18
J0808+4342	22207 ± 543	0.531 ± 0.036	28	10.3 ± 3.7	10.1 ± 3.6	12.5 ± 4.5	12.8 ± 4.1
J0823+3111	19764 ± 366	0.485 ± 0.026	44	18.9 ± 2.7	18.5 ± 2.7	14.6 ± 2.2	14.8 ± 2.3
J0842+3338	10779 ± 182	0.589 ± 0.044	481	3.8 ± 1.2	3.7 ± 1.1	6.5 ± 2.5	6.7 ± 2.5
J0851+2713	13220 ± 684	0.757 ± 0.088	410	0.9 ± 0.2	1.0 ± 0.2	1.1 ± 0.2	1.1 ± 2.3
J1121+3200	13835 ± 868	0.734 ± 0.077	341	0.9 ± 0.2	0.9 ± 0.2	1.1 ± 0.2	1.1 ± 0.2
J1133+4711	16714 ± 1820	1.228 ± 0.040	956	1.0 ± 0.0	1.0 ± 0.0	1.0 ± 0.0	1.0 ± 0.0
J1138+5156	16588 ± 928	0.854 ± 0.098	276	0.5 ± 0.1	0.5 ± 0.1	0.5 ± 0.1	0.5 ± 0.0
J1147+4758	15217 ± 1024	0.684 ± 0.100	228	2.1 ± 0.8	2.1 ± 0.7	2.9 ± 1.2	2.9 ± 1.2
J1157+4701	30584 ± 898	0.581 ± 0.083	9	6.2 ± 2.7	6.2 ± 2.7	8.5 ± 3.7	8.2 ± 3.8
J1238+3325	28129 ± 774	0.628 ± 0.065	12	2.4 ± 1.0	2.3 ± 1.0	3.9 ± 1.7	3.9 ± 1.7
J1303+3338	10635 ± 363	$0.664^{+0.197}_{-0.162}$	649	5.2 ± 2.2	5.1 ± 2.1	5.6 ± 2.5	6.1 ± 2.6
J1345+4001	17819 ± 436	0.569 ± 0.043	107	5.2 ± 1.8	5.1 ± 1.8	9.1 ± 3.6	9.0 ± 3.7
J1430+2839	11050 ± 327	0.657 ± 0.095	525	3.1 ± 1.1	3.0 ± 1.1	4.1 ± 1.6	4.1 ± 1.7
J2328+2138	18918 ± 514	0.571 ± 0.046	70	5.1 ± 1.9	5.1 ± 1.8	8.6 ± 3.5	8.8 ± 3.6

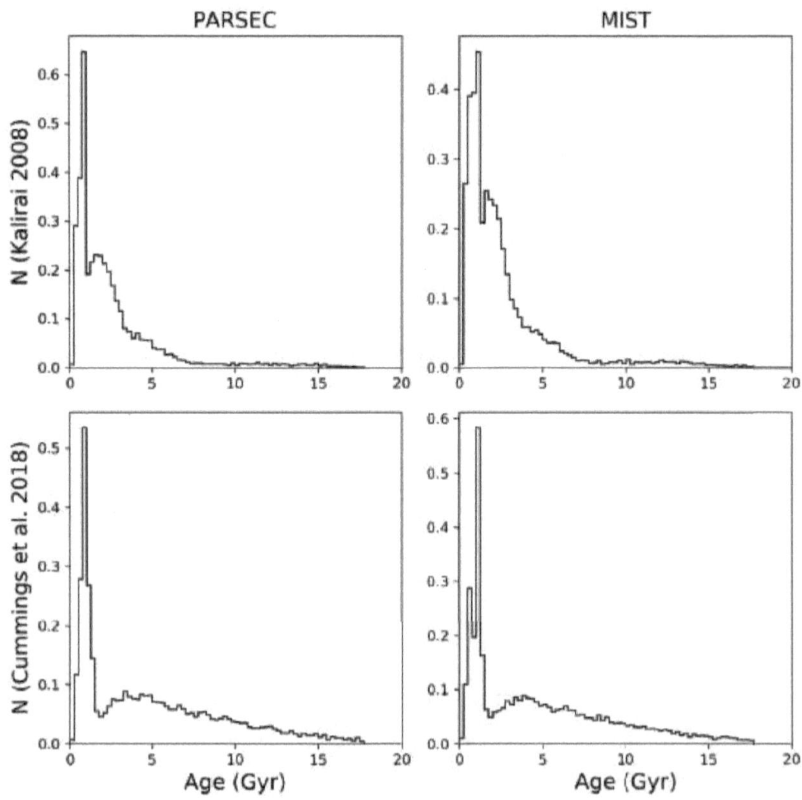

Figure 4.13 Age distribution for our 14 white dwarfs with spectroscopically confirmed masses. Each object is plotted as a Gaussian with a mean equal to the age presented in Table 4.3 and standard deviation equal to the error.

IFMR. In particular, many of the total ages of our objects are on the order of $1-2$ Gyr. Only six having ages consistent with being > 10 Gyr, although one (J0823+3111) has an age much larger than the age of the Universe.

Ages using Kalirai (2012)

The IFMR is typically calculated empirically using star clusters as the constant age of the stars means that the initial and final masses can both be accurately measured. Observationally, however, calibrating the IFMR at low initial mass is difficult as the old clusters are typically located at large distances from the Sun, making even the hot white dwarfs in the cluster very faint. This led Kalirai (2012) to construct a relationship between the final white dwarf mass using the mean mass of four field halo white dwarfs and the mean mass of six white dwarfs from M4,

$$\log[\text{Age (Gyr)}] = (\log[M_{\text{final}}/M_\odot + 0.270] - 0.201)/0.272. \tag{4.1}$$

Applying this relation to field stars comes with certain caveats. Most significantly, the metallicity may vary. This may affect the progenitor age such that the resulting white dwarf does not fall within the calibrated mass range.

We apply this relation to our sample, noting that many of our objects do not fall within this calibrated relation, and therefore incur a systematic bias. Indeed, Figure 4.14 shows that the ages of our higher mass objects increase significantly, essentially forcing them into an age compatible with the traditional view of a 12-14 Gyr halo.

4.3.5   Uncertainties and Biases

As Table 4.2 shows, many of our objects have mass uncertainties of 0.05-0.1 $M_\odot$. At the high-mass end, this translates to a fairly small uncertainty on the age — the difference in main-sequence age between a $5\,M_\odot$ and $6\,M_\odot$ star is only $\sim30\,$Myr —, whereas, at low initial mass this can lead to uncertainties on the order of the age of the Universe. Take, for example, J1147+4758, which has a mass of $0.684\pm0.1\,M_\odot$. Using the Cummings et al. (2018) IFMR, this object descended from a star with a mass of 1.1-3.2 $M_\odot$, which have main-sequence lifetimes of $0.3 - 4.5\,$Gyr.

Interestingly, it seems that many of the objects with large uncertainties also have higher masses than expected for an old halo population. In Figure 4.15 we check for bias in the fitting routine by plotting the SNR per resolution element against the white dwarf mass. The SNR is calculated in two places: (1) at H_β (top) and

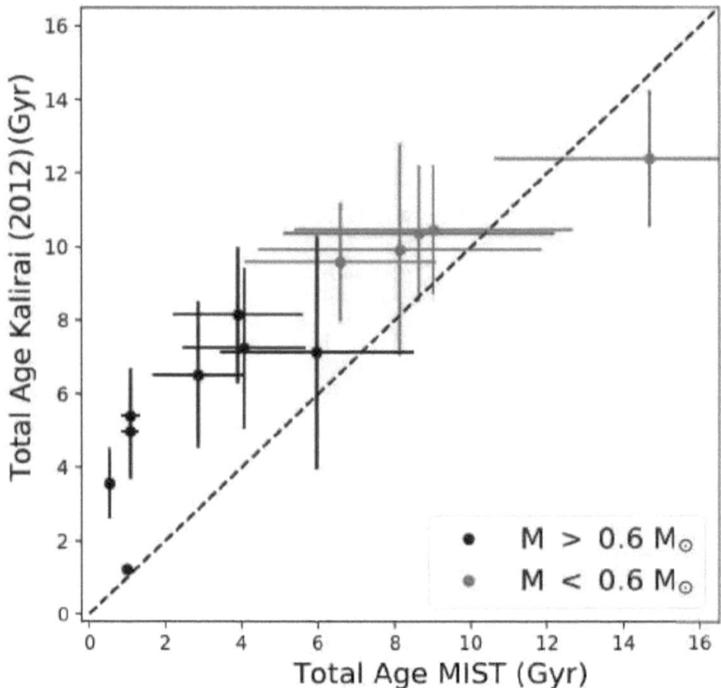

Figure 4.14 Total ages using Equation 4.1, with objects of mass greater (black) less than (red) $0.6\,M_\odot$ highlighted compared to the ages obtained using the Cummings et al. (2018) IFMR and MIST isochrones.

(2) at H_8 (bottom). Both plots suggest a systematic bias, or at the very least an underestimation in the mass uncertainty, in the fitting routine, as higher SNR values, particularly at H_8, have lower white dwarf masses. Figure 4.15 also shows that all of the high-mass objects do not pass the criteria we set out to achieve for the observations as they have SNR values below 15 at H_8.

Kepler et al. (2006) used Gemini spectra of white dwarfs near 12,000 K, the location of the maximum Balmer line equivalent width, and showed that for a given object, higher SNR data at H_8 returned lower white dwarf masses. We attempt to test this by fitting combinations of sub-exposures for objects with high SNR and low mass to see if their mass decreases with increased SNR. The results are shown in Figure 4.16, where, for each object, one sub-exposure has been added per increase in SNR. The results are inconclusive, as some objects show an increase in mass as SNR increases and some show a decrease. Unfortunately, many of the sub-exposures have a relatively high SNR, so only J0808+4342, J0842+3338, J2328+2138 had single exposures below our SNR threshold of 15. These objects show a 10% decrease, a 7% increase, and a 9% increase respectively. The issues may lie in the flux calibration, and this potential bias requires more careful considerations with larger sample sizes.

4.4 Analyzing our Halo Sample

4.4.1 Halo Membership

Given that half of our sample have masses consistent with a thin disk origin, it is useful to first check whether their kinematics could be consistent with such a population. We use the dwarf/giant catalogue presented in Thomas et al. (2019) combined with radial velocities from LAMOST (Cui et al., 2012; Zhao et al., 2012) as a reference catalogue for other disk and halo stars. We use this information to calculate the 3-D velocities in spherical coordinates and present them in Figure 4.17 as the black contours. The figure shows the disk centered at $(v_r, v_\phi) = (0, 200)$, with the extended contours centered at (0,0) representing the halo population. Specifically, the green circle represents a classical non-rotating population and the red dashed oval represents a radially-biased component recently dubbed the "Gaia-Sausage" (Fattahi et al., 2019).

Over-plotted on Figure 4.17 are our white dwarfs with distances calculated using the *Gaia* parallaxes (top) and the spectroscopic distances (bottom). These distances are calculated using the absolute magnitude calculated as part of the spectroscopic

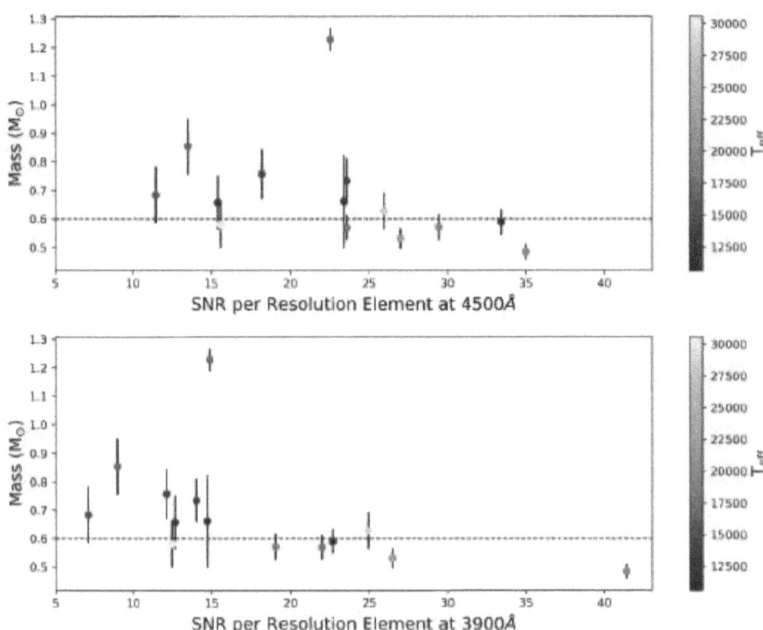

Figure 4.15 White dwarf mass vs. signal-to-noise (SNR), colour-coded by temperature. The top panel shows the SNR calculated at H_β, while the bottom plot shows the SNR calculated at H_8. The lower plot, in particular, shows a systematic decline in mass as a function of increasing SNR, suggesting a potential bias in the fitting routine.

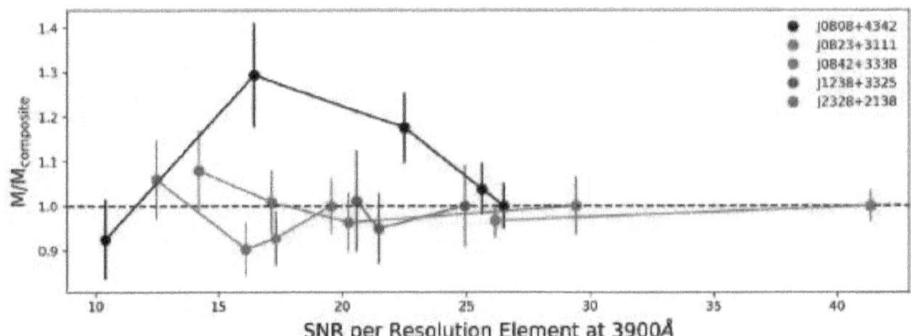

Figure 4.16 Mass determinations, as a fraction of the composite mass presented in Table 4.2, for low mass sub-exposures.

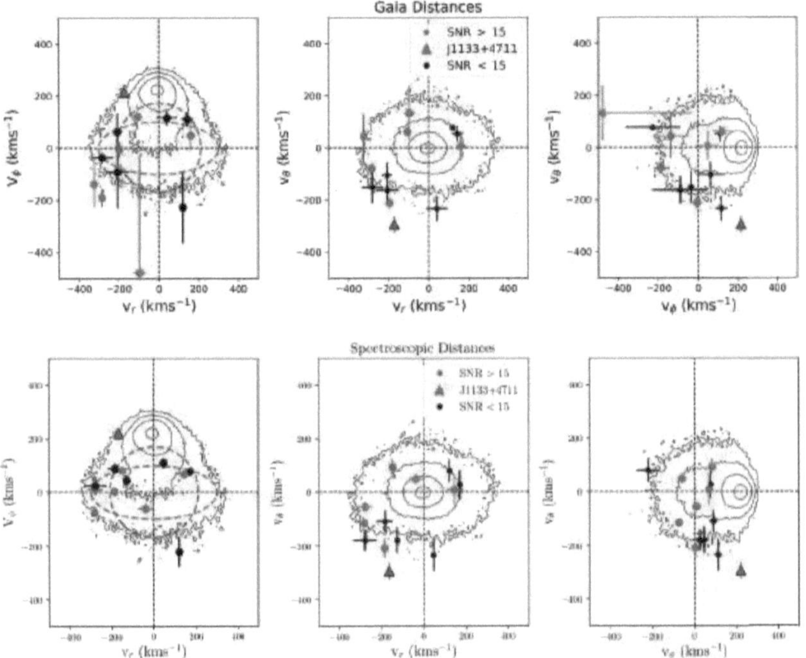

Figure 4.17 Spherical velocity distribution using distances calculated using parallax measurements (top) and the spectroscopically determined absolute magnitude (bottom). We also show a classical halo (green circle) and a more flattened halo (red oval) which is attributed to the Gaia-Enceladus merger event (Fattahi et al., 2019).

fit combined with extinction calculated from the Schlegel et al. (1998) dust maps. Figure 4.17 shows that the objects with SNR > 15 at H_8 do not populate a unique location within these diagrams, suggesting that both samples are on similar orbits. This means that the white dwarfs, with a possible exception of our high mass WD J1133+4711 (red triangle), have velocities inconsistent with a disk origin as they are on highly radial orbits with minimal rotation.

To emphasize this point, we simulated our selection function using our white dwarf population synthesis code from Chapter 3. The velocity distributions were taken from Robin et al. (2017), who derived them by comparing the Besançon model to kinematic data from RAVE and *Gaia* DR1. The resulting 1-, 2-, and 3-σ velocity ellipsoids of the thin disk, thick disk, and halo are shown in Figure 4.18. The resulting model values show that all of our white dwarfs lie beyond the 3-σ thin disk region, with the majority also falling outside of the 3-σ thick disk ellipsoid as well.

The model can also simulate the number of synthetic white dwarfs that would pass our selection criteria presented in Section 4.2. To calculate the probability of a thin disk or thick disk object of lying below the 200 kms^{-1} we ran 10,000 simulations using the parameters presented in Chapter 3. The resulting fraction of thin disk objects which would pass our selection method is 3×10^{-4} %, while 0.07 % of thick disk objects would be selected. Scaling for the mass differences, and given that we observed 13 objects, the expected number of thin and thick disk objects are 0.07 and 1.5 respectively.

It is thus unlikely that all of these objects are simply interloping young thin disk objects and that other factors are responsible for these young objects with halo-like velocities. However, given the large rotational velocity of our highest mass object, J1133+4711, this may be a truly young white dwarf on an extreme disk orbit.

We note that we do expect some thick disk objects to pass our selection criteria, however, given the observed metallicity of thick disk stars, most models of the thick disk assume ages of $\sim$10 Gyr (see, e.g, Snaith et al., 2015; Kilic et al., 2017), which would mean that the currently observable hot white dwarfs descending from this population would have masses $\sim$0.58 $M_\odot$, lower than the high-mass objects we observe.

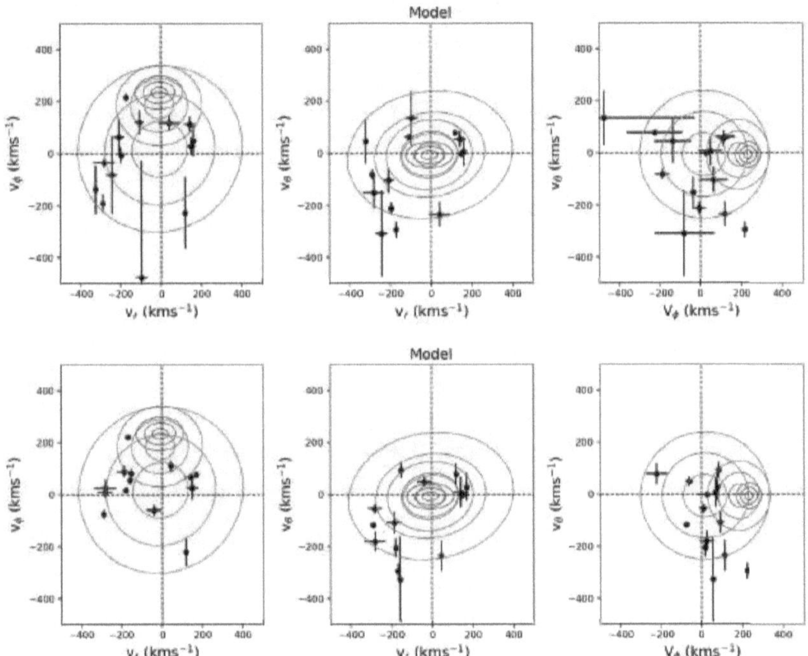

Figure 4.18 Model velocity ellipsoids using the synthetic white dwarf population within the CFIS footprint from Fantin et al. (2019). The 1-, 2-, and 3-σ velocity ellipsoids for a thin disk (blue), thick disk (green), and halo (red) are shown. The white dwarfs are plotted using distances calculated from the *Gaia* parallaxes (top) and spectroscopic absolute magnitude (bottom) as in Figure 4.17.

4.4.2   Class 1: Sample with Accurate Masses

As Section 4.3.5 shows, there likely exists a bias in that the fitting routine prefers higher masses at low SNR unless the mass is quite high. Therefore, we split our sample into three categories: (1) the high SNR objects for which we trust the mass estimates (J0808+4342, J0823+3111, J0842+3338, J1238+3325, J1345+4001, J2328+2138), (2) J1133+4711, which has an SNR below our threshold but has a small uncertainty due to its extreme mass, which is confirmed by the photometric method, and (3) the remaining objects with an SNR at 3900Å below 15. We will perform an analysis on the first category here, before discussing the remaining two in the following section.

In Figure 4.19 we present the age histogram for the Class 1 objects. Removing the low SNR objects increases the mean age, while also removing the majority of young objects seen in Figure 4.13. This is particularly true when viewing the more recent IFMR from Cummings et al. (2018), which contains a majority of objects with ages at, or above, 10 Gyr.

If we only consider the objects which passed our original criteria, the mean mass becomes 0.561 ± 0.007, suggesting the halo has a median age of 9.3 ± 1.4 Gyr using the Cummings et al. (2018) IFMR and MIST isochrones, or 10.8 ± 0.6 Gyr using the relation from Kalirai (2012). These results are consistent within the uncertainties with the result of Kalirai (2012), who derived an age of 11.7 ± 0.7 Gyr, as well as the age of 10.9 ± 0.4 from Kilic et al. (2019).

This sample can also be used to calculate the dispersion in age. In order to disentangle the contribution by measurement errors to the measured dispersion we use the method of Martin et al. (2007) and Collins et al. (2013). This method maximizes the likelihood function,

$$ML(\mu, \sigma_\tau) = \sum_{i=1}^{N} \frac{1}{\sigma_{\text{tot}}} \exp\left[-\frac{1}{2} \left(\frac{\mu - \tau_i}{\sigma_{\text{tot}}} \right)^2 \right] \tag{4.2}$$

where τ_i are the individual ages of each star, μ is the mean age of the population, and $\sigma_{\text{tot}} = \sqrt{\sigma_\tau^2 + \sigma_i^2}$ is the sum of the dispersion from measurement errors, σ_i, and the intrinsic dispersion, σ_τ.

We use Emcee (Foreman-Mackey et al., 2013) to maximize the likelihood function and to estimate the uncertainties associated with the mean and dispersion. The results of simulations with one thousand walkers each taking one million steps can be seen in Figure 4.20.

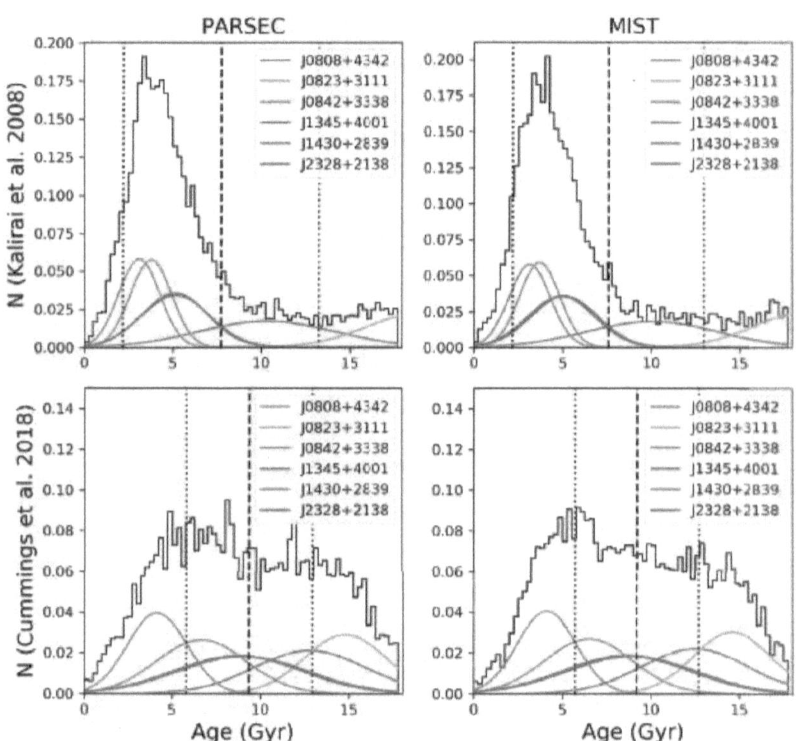

Figure 4.19 Age histograms for objects belonging to Class 1 (see 4.4.2 for discussion). The Cummings et al. (2018) IFMR returns lower initial masses, resulting in larger overall ages. Also shown are the mean (solid black line), and 1σ values (dashed black lines).

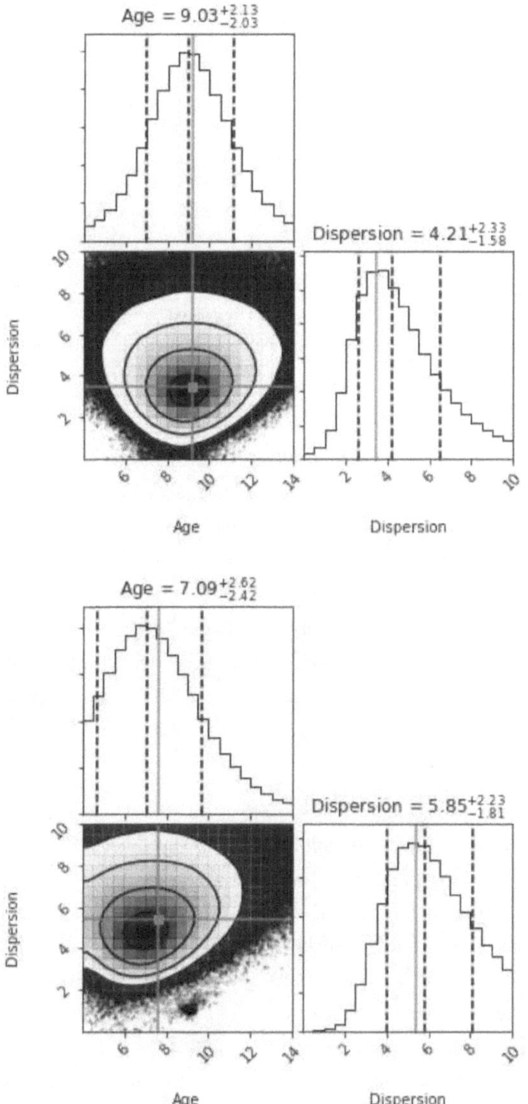

Figure 4.20 Corner plots for the mean age and intrinsic dispersion in ages for Class 1 objects. (Top): Using the MIST isochrones and Cummings et al. (2018) IFMR. (Bottom:) Using MIST isochrones and Kalirai et al. (2008) IFMR.

Given that changing isochrones did not have a large effect on the total age the simulations use the MIST isochrones and only show a difference based on the choice of IFMR. Using the most up-to-date IFMR from Cummings et al. (2018), a mean age of $9.03^{+2.13}_{-2.03}$ Gyr is measured. The intrinsic dispersion was determined to be $4.21^{+2.33}_{-1.58}$ Gyr.

This suggests a more extended star formation history than is typically assumed for this population ($\sim$1 Gyr. see e.g, Reid, 2005; Kilic et al., 2017). Recent evidence, however, has shown that the inner halo is more complex and likely dominated by a major accretion event that occurred around 10 Gyr ago (Helmi et al., 2018; Belokurov et al., 2018), allowing for a more extended star formation history in the progenitor. Our ages are also consistent with this event, suggesting that these white dwarfs may have formed as a result of such a merger, or even in the progenitor itself.

4.4.3   High Mass Halo White Dwarfs?

Given the inconclusive results from Section 4.3.5 regarding our low SNR objects in sample (3), we consider four scenarios for which high-mass white dwarfs on halo-like orbits could occur. The first three scenarios involve Galactic processes: (1) these objects are indeed thin disk objects but have been perturbed somehow onto halo-like orbits (2) these objects may be the end-products of double degenerate mergers; or (3) these objects are indeed old but they were able to prolong their lifetimes through binary interactions. Finally, in a fourth scenario, we consider whether these objects could have been recently accreted from an external galaxy having an extended star-formation history.

Scenario I: Ejected Thin Disk Objects

The first scenario has been proposed to explain the hypothesis of Oppenheimer et al. (2001), who selected high-velocity white dwarfs and claimed that these objects could be a significant contributor to the unseen matter in the halo. Davies et al. (2002) proposed that thin disk stars could achieve high tangential velocities if they were ejected from a binary system due to a more massive companion experiencing a type II supernova. The star would then evolve, as usual, into a white dwarf while maintaining the high-velocity. The authors simulate a binary interaction for which a 12 $M_{\odot}$ primary star undergoes a supernova resulting in the secondary star being ejected, and produce a relationship between the secondary mass and the kick velocity (see their Figure 3).

To test this hypothesis we select thin disk stars within our white dwarf population synthesis model and impose a kick velocity, selected from their derived relationship to their original velocities. We then apply the same selection function as presented in Section 4.2 to determine the fraction of ejected objects which would fall within our selected region of the reduced proper motion diagram. The total fraction of thin disk objects which received this kick velocity and passed our selection function is $0.19 \pm 0.01\,\%$. Thus, for a sample of nearly 30,000 white dwarfs, of which 80% are likely members of the thin disk, if all of them received a velocity kick we would detect on the order of 50 objects, which can be seen in Figure 4.21. In order to explain our set of high-mass objects we would require that nearly 10% of all thin disk white dwarfs be ejected during their lifetime, a number that is unrealistic given the lack of high-mass stars produced via the initial-mass function that can produce a type II supernova.

Furthermore, given that the direction of the kick velocities is likely random, and that the objects are initially rotating in the disk, we expect for this sample of objects to continue to rotate but with a larger velocity dispersion. Our sample of low SNR halo objects does not show any net rotation, and every object lags behind the local standard of rest. This comparison is highlighted in Figure 4.21, where many of the objects which received a kick do not lie in the same region of velocity space as our sample. Thus, we conclude that while objects from the thin disk can be kicked out into the halo, the discrepancy between the velocity distributions seen in the model when compared to our data makes it unlikely that our samples were drawn from the same distribution. To test this, we apply a 2-d Kolmogorov-Smirnov test to our (v_r, v_ϕ) data using the method presented in Fasano & Franceschini (1987). The test returns a p-value of 8.8×10^{-5}, suggesting that the velocities are not selected from the same distribution. We thus conclude that this process alone cannot account for the high-mass white dwarfs seen in our sample.

This scenario may explain J1133+4711, however, as it has a large radial velocity component whilst maintaining a large rotational velocity as well. It also lies close to the 3σ contour for the thin disk, suggesting that if it started with a large velocity to begin with, even a minimal kick could allow it to enter a more radial orbit.

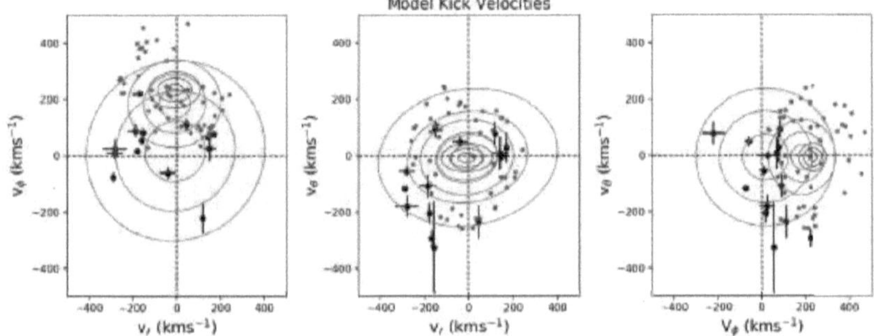

Figure 4.21 Model velocity ellipsoids using the synthetic white dwarf population within the CFIS footprint from Fantin et al. (2019). The 1-, 2-, and 3-σ velocity ellipsoids for a thin disk (blue), thick disk (green), and halo (red) are shown. The white dwarfs are plotted using distances calculated from spectroscopic absolute magnitude as in Figure 4.17. White dwarfs which originated in the thin disk and received a kick velocity are plotted in blue.

Scenario II: White Dwarf Mergers

When calculating the age using isochrones, it is assumed that the object evolved as a single white dwarf without any external influence. Recent observations have shown that the merger of two white dwarfs in a binary system can form a single, more massive white dwarf, with the new object having a mass roughly equivalent to the combined mass of the two stars, although some mass loss may occur. Temmink et al. (2020) found that the resulting merger products have masses higher than white dwarfs formed through single-star evolution, with a mean of $0.71\,M_\odot$.

The merged white dwarf scenario has been used to explain the discrepancy between the binary fraction of main-sequence stars ($\sim50\%$) and the local ($<25\,\mathrm{pc}$) white dwarf sample ($\sim26\%$ Holberg et al., 2016; Toonen et al., 2017; Kilic et al., 2018). Furthermore, there exists a high-mass 'bump' in the observed mass distribution (see, e.g, Figure 1.4 and Liebert et al., 2005; Kleinman et al., 2013; Kepler et al., 2019), which can be explained by a double degenerate merger rate of $14-25\%$. There are, however, other explanations for this high-mass bump, which include a flattening of the IFMR (El-Badry et al., 2018; Cummings et al., 2018). This explanation does not explain the discrepancy in the binary fraction between the local main-sequence stars and the white dwarfs.

Scenario III: Descendants of Blue Stragglers

Figure 4.12 shows that many of the objects have progenitor masses between 1.5 and $2\,M_\odot$, which are typical of A-type stars. Despite having main-sequence lifetimes on the order of 1 Gyr (Hurley et al., 2000; Bressan et al., 2012; Choi et al., 2016), these stars are still seen in old populations like globular clusters (see, e.g, Harris, 1993; Knigge, 2015; Parada et al., 2016b), as well as in the thick disk and halo of the Milky Way (Santucci et al., 2015) and are called blue stragglers (BSs). These objects are thought to form as a result of mass-transfer binaries or direct mergers of two stars (Stryker, 1993). The lifetimes have been estimated to be $200-300\,\mathrm{Myr}$ once formed, and their post-main-sequence evolution is the same as a star with the same initial masses (Parada et al., 2016b). Thus, halo BSs would create higher mass white dwarfs with kinematics consistent with those seen in our sample.

To study this scenario, we require a sample of metal-poor, main-sequence, A-type stars in the solar neighbourhood on halo-like orbits. Preston et al. (1994) used *UBV* photometry to select blue stars with [Fe/H]<-1.0, (dubbed blue metal-poor;

BMP) stars, within 2 kpc of the Sun and found a space density of 450^{+300}_{-150} kpc^{-3}. Follow-up spectroscopy presented in Wilhelm et al. (1999), however, showed that a large fraction of these objects ($\sim$50%) were more metal-rich ([Fe/H]$>$ -1.0) than previously believed, suggesting that many of these objects may not be from an old population.

A sub-sample of the BMP stars from Preston et al. (1994) were monitored by Preston & Sneden (2000) to test their binarity through radial velocity variations. Their results suggest that more than half of these objects were in a binary system, suggesting that they were indeed blue stragglers formed through mass accretion. Thus, if half of the original sample were true blue stragglers, and half of these were indeed metal-poor as suggested by Wilhelm et al. (1999), the space density of these blue stragglers would be $1.1^{+0.75}_{-0.38}\times10^{-7}$ pc^{-3}.

The space density for young, hot, halo objects ($T_{\text{eff}} > 10,000\,$K) studied in this work has been calculated to be 1.2×10^{-6} pc^{-3} using the luminosity function of Munn et al. (2017), a factor of 10 larger than the space density of halo blue stragglers. This is likely an upper limit, as many of these BMPs may not be part of the in-situ halo given their rotational velocity of 128 kms^{-1} presented by Preston et al. (1994), which lies intermediate to the thick disk and halo. Thus, we conclude that this scenario alone can not account for the observed high mass halo white dwarfs, although it is worth considering given the recent work by Parada et al. (2016b) which showed that blue stragglers evolve off the main-sequence as if they had always been at their current mass, suggesting they will form higher mass white dwarfs. Our upper limit of 9^{+6}_{-3}% contamination is also consistent with the increased fraction of AGB stars they observe in 47 Tucanae that they attribute to evolved blue stragglers.

Scenario IV: Accretion or Perturbation from a Satellite

Preston et al. (1994) suggested that many of these BMP stars could be of extragalactic origin as their young ages and low metallicity could occur in smaller dwarf galaxies seen surrounding the Milky Way. If one of these objects was accreted recently it could explain the origin of such objects and their resulting white dwarfs. The issue with this scenario is that, given the inferred progenitor masses, these objects were formed within the past 2 Gyr regardless of the progenitor metallicity. Combining this with the large fraction of our sample means that we would require a large merger event within the past few Gyr, which to our knowledge has not been reported within the literature (Naab & Ostriker, 2017; Ruiz-Lara et al., 2020).

Much of the recent focus has been on the Gaia-Enceladus merger event (Helmi et al., 2018; Belokurov et al., 2018), which is thought to be the last major merger event in the Milky Way. This event, however, is thought to have occurred more than 10 Gyr ago, and could be used to explain the formation of the thick disk (Helmi et al., 2018). Thus, a major merger could explain the formation of our low-mass objects, however, there is a lack of evidence for a major merging event over the last few Gyr (Naab & Ostriker, 2017; Ruiz-Lara et al., 2020).

There is, however, evidence for interactions with more minor satellites over the past few Gyr, including the Sagittarius dwarf galaxy. Ruiz-Lara et al. (2020) studied the local 2 kpc bubble and found signatures of an increased star-formation rate at three separate epochs: 5.7 Gyr, 1.9 Gyr and 1.0 Gyr, the last two of which are consistent with the ages derived in Section 4.3. These bursts of star formation coincide with the proposed pericentre passages of the Sagittarius dwarf galaxy and are visible in both the thin and thick components of the Milky Way. Ruiz-Lara et al. (2020) conclude that these episodes could be explained by the interaction between the Milky Way disk and Sagittarius, however, the exact physical mechanism has yet to be explained. Thus, the high-mass stars formed in the thick disk during one of these episodes could form the high-mass white dwarfs observed in our sample, although more studies are needed to determine the orbits of the objects formed during these events.

4.5 Summary and Conclusions

This chapter has presented follow-up spectroscopy for 18 candidate halo white dwarfs performed at the twin 8-m Gemini Observatories. Our results are as follows:

- We observe 13 DA, 2 DC, 1 DZ, and 2 peculiar white dwarfs in our sample, representing a large variation in types relative to previous samples.

- The discovery of two peculiar, but rare, white dwarfs reinforces the conclusion of Raddi et al. (2019) who suggested that these objects could form as a result of a peculiar class of supernovae occurring in binary systems that ejects the white dwarf, thus increasing its velocity.

- For the sample of DA white dwarfs, we calculate their temperature and mass by fitting either the Balmer series. The results show a peak at $0.58\,M_\odot$ with a number of outliers.

- We convert the white dwarf mass to a total age using the initial-to-final mass relation of Kalirai et al. (2008) and Cummings et al. (2018) in combination with MIST and PARSEC stellar isochrones. The ages show a large spread, which is typically not seen in a halo sample.

- We investigate whether the model fitting routine imposes a systematic bias towards higher white dwarf masses at lower signal-to-noise and find that the model predicts higher masses for objects with lower signal-to-noise ratios at H_8. We attempt to study this potential bias in more detail by fitting a combination of sub-exposures for the low-mass, high SNR objects, however, the results do not show a decrease or increase in mass as the SNR increases.

- Given the potential of bias in the sample, we split the white dwarfs into three classes. (1) Those with SNR > 15 at H_8 as recommended by Kepler et al. (2006), (2) J1133+4711 which has a mass of 1.22 $M_\odot$ in both the spectroscopic and photometric methods, and (3) the remaining objects with SNR < 15.

- We find that the high SNR objects have a mean age of 9.3 ± 1.4 Gyr using the Cummings et al. (2018) IFMR and MIST isochrones, or 10.8 ± 0.6 Gyr using the relation from Kalirai (2012). These results are consistent with previous studies.

- We determine the intrinsic dispersion in the halo ages for Class (1) to be $4.21^{+2.33}_{-1.58}$ Gyr using a maximum likelihood method. This suggests a more extended star formation history for the halo than is typically assumed.

- We investigate four scenarios that would produce high-mass white dwarfs on halo-like orbits, such as those in class (2) and (3). Using our Milky Way white dwarf population synthesis code we find that ejected thin disk objects alone can not explain the sample, however, a combination of white dwarf mergers, halo blue stragglers, or increased star formation as a result of an interaction with the Sagittarius dwarf galaxy could explain this sample.

This work highlights the need for high signal-to-noise data, particularly at H_8, when measuring white dwarf masses as low-SNR data may induce a bias in the resulting white dwarf masses. This is particularly true around 12,000 K where the Balmer lines are at their maximum equivalent widths. We show that the average

age of low SNR objects is between 1 and 3 Gyr, whereas when we exclude the objects with SNR below the recommendation from Kepler et al. (2006), we find the age of the inner halo to be $9.3\pm1.4\,$Gyr using the Cummings et al. (2018) IFMR and MIST isochrones, or $10.8\pm0.6\,$Gyr using the relation from Kalirai (2012), which are consistent with previous results.

We note, however, that we can not rule out these objects as truly being high-mass white dwarfs belonging to the Galactic halo. Blue stragglers, double degenerate mergers, or enhanced star formation in the thick disk due to an interaction with a dwarf galaxy can all produce high-mass white dwarfs with halo-like velocities. Future large spectroscopic surveys such as WEAVE (Dalton et al., 2012), 4MOST (de Jong et al., 2012), and the Mauna Kea Spectroscopic Explorer (The MSE Science Team et al., 2019) will be needed to increase the spectroscopic sample of halo white dwarfs to accurately determine their mass distribution.

This work has also reiterated the importance of the IFMR. In particular, much work still needs to be done at the low-mass end where halo stars are currently producing white dwarfs. Future large observatories such as the Mauna Kea Spectroscopic Observatory and the extremely large telescopes (the Thirty Meter Telescope, the Giant Magellan Telescope, and the Extremely Large Telescope) will be needed to acquire a large sample of spectroscopic masses in globular clusters given their large distances and faint populations. In addition to the ground-based observatories, recent work by Gentile Fusillo et al. (2019) suggests that the James Webb Space Telescope can also be used by targeting the Paschen series in the near-infrared. Combining these observatories with the previously mentioned spectroscopic surveys will provide an unprecedented look into the formation and evolution of the Galactic inner halo.

Chapter 5

The Future of White Dwarf Science in the Upcoming Decade

5.1 White Dwarfs and Wide-Field Surveys

The upcoming decade will see the construction of several powerful new observatories with wide-ranging science cases. Previous generation sky surveys, from the Palomar Observatory Sky Survey (Minkowski & Abell, 1963; Reid et al., 1991), to more re-cent surveys like the SDSS (York et al., 2000) and *Gaia* (Gaia Collaboration et al., 2018b), have established that an incredible breadth of research can be undertaken with publicly-accessible, wide-field survey data. This includes the study of white dwarfs — the end stage of stellar evolution for the vast majority of stars.

White dwarfs have been used, among other things, to test fundamental physics (see, e.g, Winget & Kepler, 2008; Hansen et al., 2015), to study stellar populations of all ages (see, e.g, Richer et al., 1997; Bedin et al., 2009; Salaris & Bedin, 2018), and to explore Milky Way formation and evolution (Winget et al., 1987; Rowell, 2013; Kilic et al., 2017; Fantin et al., 2019). This widespread utility reflects the fact that white dwarfs represent extreme physical environments and yet are ubiquitous throughout the Milky Way.

Despite their importance, detecting these inherently faint objects can be a challenge. Historically, white dwarfs have been discovered in large numbers in wide-field, multi-epoch, optical surveys as either high proper motion objects (e.g, Luyten, 1979; Rowell & Hambly, 2011), or as star-like objects with a UV excess Green et al. (1986). Such surveys have uncovered hundreds to thousands of white dwarfs, with the majority residing within the local neighbourhood.

Spectroscopic and photometric samples of white dwarfs increased by orders of magnitude with the beginning of the SDSS (York et al., 2000; Kleinman et al., 2004). Photometrically, white dwarf samples on the order of $\sim10^4$ have been generated using SDSS data as a second epoch to derive proper motions (Harris et al., 2006; Munn et al., 2017), allowing for an accurate determination of the local luminosity function. Furthermore, the spectroscopic sample currently stands at more than 20,000 objects as of SDSS Data Release 14 (Kepler et al., 2019), although this sample was acquired mainly as a result of a targeted QSO survey which imparts a significant temperature bias on the sample (see e.g, Kleinman et al., 2013).

White dwarfs can also be identified using parallaxes as their intrinsic faintness allows them to be differentiated from other stars with equivalent colours. With the release of *Gaia* DR2, the number of white dwarf candidates has increased by another order of magnitude, with recent catalogues containing on the order of 10^5 white dwarfs (Jiménez-Esteban et al., 2018; Gentile Fusillo et al., 2019).

Because the magnitude limit of the *Gaia* catalogue is only $G \sim 20.5$, future surveys are poised to further increase the sample of known white dwarfs by many more orders of magnitude. In this Chapter, we examine the scope of white dwarf science that can be performed using some major ground- and space-based surveys that will be carried out later this decade — the Legacy Survey for Space and Time (LSST), Euclid, and the Roman Space Telescope. We also consider the proposed Cosmological Advanced Survey Telescope for Optical and uv Research (CASTOR), which would provide UV/blue-optical imaging that is complementary to these other facilities.

We begin by describing the key aspects of each survey in §5.2 and then briefly describe our model in §5.3. The simulated white dwarf samples expected within each survey are discussed in §5.4 before we discuss in §5.5.1 the various white dwarf selection options using the LSST. In §5.5.2, we highlight some notable white dwarf research opportunities that would be possible in the Roman Space Telescope's High Latitude Survey, a ~2000 deg^2 region that will be included in all surveys. We finish with an analysis of some other, representative white dwarf science cases in §5.6 before summarizing in §5.7.

5.2   The Surveys

In this section, we describe four planned surveys with nominal start times in this decade. We highlight the relevant capabilities including spatial coverage, photometric depths, and whether any astrometric measurements will be performed.

5.2.1   The Legacy Survey for Space and Time (LSST)

The Simonyi Survey Telescope (formerly the Large Synoptic Survey Telescope), part of the newly named Vera C. Rubin Observatory, is an upcoming ground-based observatory in the Southern Hemisphere. With an 8.4m primary mirror and nearly 10 deg^2 field of view, the observatory will efficiently image much of the Southern sky for a period of 10 years with a focus on time-domain astronomy. The resulting survey — The Legacy Survey of Space and Time — will produce precise photometry and astrometry nearly 4 magnitudes deeper than the current *Gaia* mission over a large swath of the Southern sky (LSST Science Collaboration et al., 2009; Ivezić et al., 2019).

While the survey itself consists of a few different regions of the sky, in this chapter we focus on the main survey, dubbed the "wide-fast-deep" (WFD) survey. This survey will cover the declination range $-65° \leq \delta \leq +5°$ while avoiding a region of the Galactic Plane (see the top panel of Figure 5.1). The survey will produce a single visit 5σ point-source r-band magnitude limit of 24.5, culminating with co-added depths of $r \sim 27.5$ at the completion of the survey (LSST Science Collaboration et al., 2009).

The survey will be performed in six optical bands, *ugrizy*, as shown in the bottom panel of Figure 5.1. The survey will also be performed under very good seeing conditions, with median conditions between $0.55''$ in the u-band to $0.46''$ in the y-band.

A key strength of the LSST is its ability to measure accurate parallax and proper motion measurements to depths previously only attainable in space-based missions. The survey achieves this by imaging each part of the sky in two 15-sec exposures, called a visit, more than 800 times per source (LSST Science Collaboration et al., 2017). This will revolutionize the field of time-domain astronomy, which includes everything from the most distant supernovae to more local stellar populations like white dwarfs.

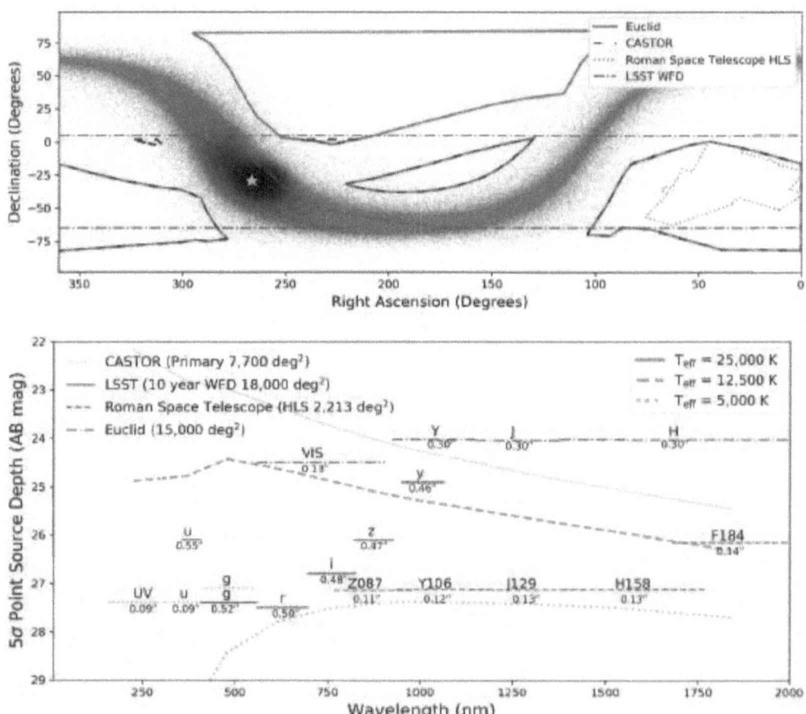

Figure 5.1 *Top:* Footprints for the various surveys in equatorial coordinates. Also plotted are model white dwarfs for reference, ignoring extinction, to highlight the location of the Galactic disk. The Galactic centre is marked with a star. *Bottom:* Magnitude limits, wavelength coverage, and median image quality for each survey. Also plotted are the SEDs for $0.6\,M_\odot$ white dwarfs of three temperatures each placed at a distance of $3{,}800\,\mathrm{pc}$: $25{,}000\,\mathrm{K}$ (solid black), $12{,}500\,\mathrm{K}$ (dashed blue), and $5{,}000\,\mathrm{K}$ (dotted red) for reference.

The Observatory is currently being constructed at Cerro Pachón in the Andes mountains of Chile, with first light expected in 2021. The main survey is expected to begin in October 2022 and last 10 years, producing more than 200 Petabytes of data in the process.

5.2.2 Euclid

The Euclid space mission (Laureijs et al., 2011) is the European Space Agency's upcoming near-infrared (NIR) survey designed to investigate dark energy through weak lensing and baryonic acoustic oscillation measurements. This will be accomplished by surveying billions of galaxies over $\sim$15,000 deg^2 of sky (Racca et al., 2016). However, like all large surveys, there are many other ancillary science cases. Most relevant for this study is the ability of NIR imaging for white dwarfs to reveal dusty debris disks and sub-stellar companions like cool M-dwarfs and brown dwarfs (Dennihy et al., 2017).

Euclid's Wide Survey will focus on high Galactic latitude fields, imaging in three NIR bands (JHK), and one wide optical band (VIS) to 5σ depths of $\sim$24 in the NIR and $\sim$25.5 in the optical. With a 1.2m primary mirror and 0.5 deg^2 field-of-view, the resulting images will have excellent resolution: i.e., between 0.1$''$ and 0.3$''$.

As Figure 5.1 shows, the main survey field (solid green) will overlap with many of the other surveys, including nearly 40% of the LSST WFD survey, nearly the entire CASTOR primary survey, and the entire Roman Space Telescope HLS. The launch time as of publication is expected to be in 2022[1], with an expected mission duration of seven years, meaning that all data will be in hand before the completion of the LSST.

5.2.3 The Nancy Grace Roman Space Telescope

The Nancy Grace Roman Space Telescope, shortened as the Roman Space Telescope, and formerly the Wide-Field InfraRed Survey Telescope (WFIRST) (Spergel et al., 2015), is an upcoming NASA mission that also aims to study dark energy. The telescope consists of a 2.4m primary mirror with a field-of-view of 0.28 deg^2. The Roman Space Telescope's High Latitude Survey (HLS) will image roughly 2,200 deg^2 of the Southern Sky, in a region that fully enclosed within the LSST WFD footprint.

The survey will include optical and NIR imaging, from $5000\,\text{Å}$ to $2\,\mu\text{m}$ to a depth of >26.5 AB mag. The image quality will be roughly $0.1''$ (see Figure 5.1). Since the imaging will be collected over the 5 year duration of the mission, absolute proper motions for many stars will be measured (Spergel et al., 2015), allowing for many ancillary Galactic science cases, including the detection of some of the coolest white dwarfs in the Galactic halo. The Roman Space Telescope is set to launch in the mid-2020s. If this time-frame is realized, then the Roman Space Telescope's data products will be available before the completion of LSST's 10-year survey.

5.2.4 The Cosmological Advanced Survey Telescope for Optical and uv Research (CASTOR)

The Cosmological Advanced Survey Telescope for Optical and uv Research (CASTOR) is being developed by the Canadian Space Agency (CSA) for a possible launch in the late 2020s. This 1m diameter, off-axis telescope would use a three-mirror anastigmat design to deliver high-resolution imaging (FWHM $\simeq 0.''15$) over an instantaneous 0.25 deg^2 field of view (Côte et al., 2012; Cote et al., 2019). Dichroics and multi-layer coatings are used to define its photometric passbands — UV (0.15–0.30 μm), u' (0.30–0.40 μm) and g (0.40–0.55 μm) — and deliver simultaneous images to three distinct focal plane arrays.

Among its various science programs, CASTOR would carry out a "primary survey" that would image the $\sim$7700 deg^2 region defined by the overlap of the LSST WFD, Euclid Wide and the Roman Space Telescope HLS footprints. This survey would reach a (5σ) point-source depth of $m_{\text{AB}} \sim 27.2$ mag. By focusing on the UV/blue-optical region, CASTOR would complement the Euclid and the Roman Space Telescope missions by expanding the available SED coverage; relative to LSST, CASTOR would add coverage in the UV region and provide space-quality resolution at the shortest wavelengths accessible from the ground.

5.3 The Model

We model each survey using the white dwarf population synthesis code presented in Chapter 2 using the star formation history presented in Chapter 3.

The model was modified to include the relevant survey parameters including the survey area and magnitude completeness corrections as presented in Figure 5.1. The completeness functions are applied as in Figure 3.3 since they allow for a smooth

Table 5.1 <u>Photometric Information</u>

Passband	$\lambda_{mean}(\text{Å})$	A_λ	Reference
lsst u	3693.2	4.145	a
lsst g	4797.3	3.237	a
lsst r	6195.8	2.273	a
lsst i	7515.3	1.684	a
lsst z	8664.4	1.323	a
lsst y	9710.3	1.088	a
CASTOR UV	2290.0	7.070	b
CASTOR u	3460.0	4.249	b
CASTOR g	4800.0	3.230	b
Roman Space Telescope R062	6200.0	2.374	c
Roman Space Telescope Z087	8685.0	1.436	c
Roman Space Telescope Y106	1059.5	1.023	c
Roman Space Telescope J129	1292.5	0.755	c
Roman Space Telescope H158	1577.0	0.541	c
Roman Space Telescope F184	1841.5	0.413	c
Euclid VIS	7250.0	1.781	c
Euclid Y	1033.0	1.004	c
Euclid J	1259.0	0.702	c
Euclid H	1686.0	0.455	c

Note. — We assume $R_V = 3.1$.
a – (Schlafly & Finkbeiner, 2011).
b – Using Equation A1 of (Schlafly & Finkbeiner, 2011) c
– Padova Isochrones

drop-off beyond the 5σ limits, and we apply a shift based on the fainter magnitude limits of each survey.

The resulting mock catalogues contain an object's position, mass, cooling age, distance, proper motion, Galactic space velocities, as well as the observed magnitude and reddening for each relevant survey. To highlight the results of our model we show the colour-colour diagrams, with photometric uncertainties, for objects in each survey in Figure 5.2. The figure illustrates the exquisite photometry that will be acquired as part of each survey.

The results also highlight the separation between pure-Hydrogen and pure-Helium atmospheres. Since the Balmer lines are located in the u- and g-bands it is no surprise to see the populations well separated in the LSST and CASTOR samples.

Furthermore, Figure 5.2 shows the density of points throughout the colour-colour diagram. Since white dwarfs cool as they age, redder colours generally indicate cooler, older, white dwarfs. It can be seen that the densest location present in the NIR surveys occur at red temperatures, indicating a larger concentration of these cool objects relative to the hot white dwarfs. In the optical/UV surveys, more uniform distributions can be seen, indicating a larger number of hot, young, white dwarfs. These samples will be explored in detail in the following section.

5.4 Survey Results

Here we present the results of each simulation while highlighting the strengths of each survey. These results present the total number of objects in each survey and selection methods are discussed in the following section.

5.4.1 Magnitude Distribution

The magnitude distributions for each survey are presented for all white dwarfs in the left-hand panel of Figure 5.3. Given the large area and extreme photometric depths that LSST will probe, it is no surprise to see this survey producing the greatest number of white dwarfs. At the single epoch depth of $r \sim 24.5$, the survey will observe on the order of 16 million white dwarfs, and, at the full stacked depth of the 10-year survey, it will observe more than 150 million white dwarfs.

Impressive samples will also result from CASTOR, the Roman Space Telescope, and Euclid. CASTOR will observe roughly 8 million white dwarfs in the ultraviolet, allowing for the most complete sample of recently formed white dwarfs, which will be

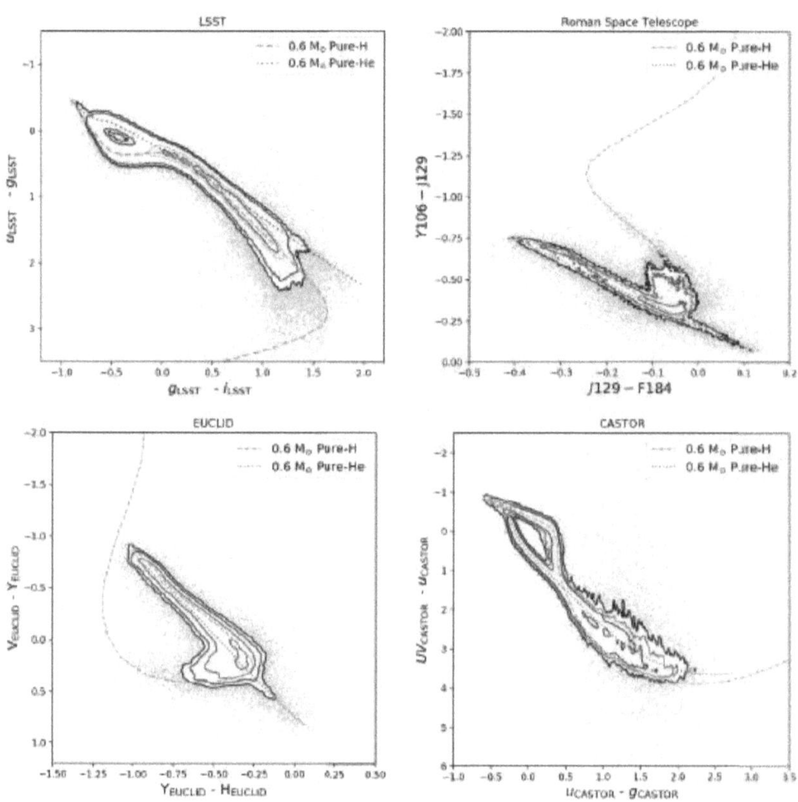

Figure 5.2 Colour-colour diagram for the four surveys with model $0.6\,M_\odot$ pure-H (solid cyan) and pure-He (dotted magenta) cooling curves over-plotted. The simulation includes the addition of photometric uncertainties.

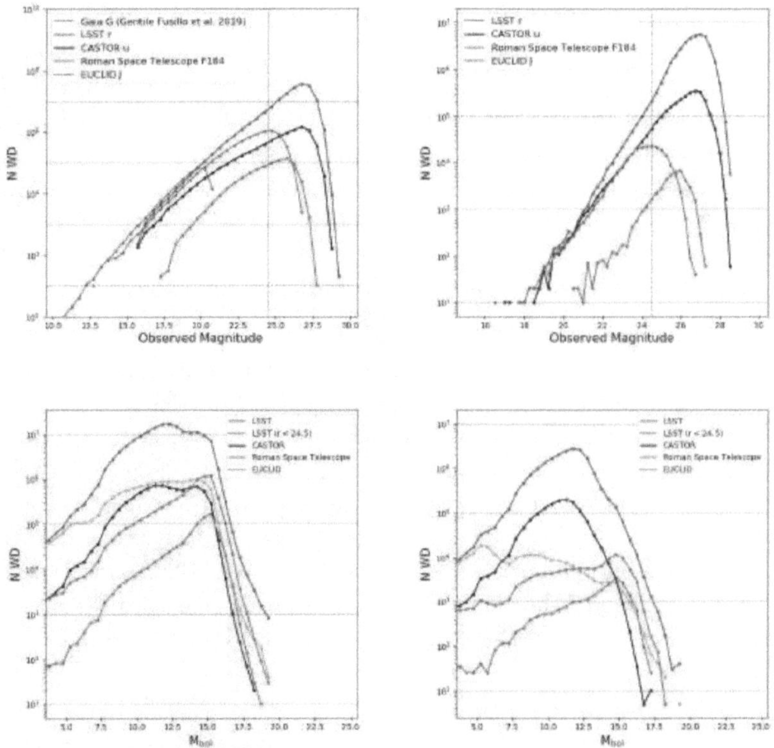

Figure 5.3 *Top:* The observed magnitude distributions for all Galactic white dwarfs (left) found within each survey, with the halo component shown in the right column. These magnitudes include the effect of interstellar dust extinction as detailed in §5.3. *Bottom:* The bolometric magnitude distribution for each survey for all white dwarfs (left) and just the halo population (right). These distributions represent the temperature distribution of the simulated white dwarfs, showing that the hot white dwarfs will be observed by the UV surveys while the cool ones will be predominately observed by the IR surveys.

important for the initial-to-final mass relation and fundamental physics like neutrino cooling (Hansen et al., 2015). The Roman Space Telescope will detect nearly a million white dwarfs in its NIR bands, which will include the coolest and ancient white dwarfs, as well as those with debris disks. Euclid will detect nearly 7 million white dwarfs in its red-optical and NIR bands over a large area covering both hemispheres.

Also highlighted in the right-hand panel of Figure 5.3 is the magnitude distributions for samples of halo white dwarfs. LSST will detect the largest sample of halo objects, with approximately 50 million to full depth and nearly half a million to the single epoch depth of $r \sim 24.5$. CASTOR's performance is also noteworthy, with nearly three million halo white dwarfs observed within the survey.

5.4.2 Bolometric Magnitude Distributions

In the bottom panels of Figure 5.3 we also show the bolometric magnitudes of the white dwarf samples. The bolometric magnitude is indicative of surface temperature, with fainter magnitudes representing cooler and older objects.

The plots highlight the relative strengths of the different surveys, particularly when it comes to halo objects. Surveys such as the Roman Space Telescope's HLS and the Euclid Wide Survey, while observing fewer overall white dwarfs, will observe more of the coolest and most ancient objects that lie beyond the turn-off in the white dwarf luminosity function (see §5.6). CASTOR and LSST will observe many of the newly formed white dwarfs in the halo and, therefore, the union of the four surveys will observe white dwarfs covering a wide range in surface temperature.

5.4.3 Colour-Magnitude Diagrams

In Figure 5.4 we show the resulting colour-magnitude diagrams for each survey, while also highlighting the model cooling tracks for $0.6 \, M_\odot$ and $1.0 \, M_\odot$ pure-Hydrogen white dwarfs. The results show that the NIR surveys, the Roman Space Telescope and Euclid, will observe a larger fraction of cool white dwarfs, while CASTOR will primarily see the hotter white dwarfs. LSST, with its broad optical wavelength coverage and large area, will observe a significant number of both young and old white dwarfs.

5.5 Survey Selection Methods

In this section, we describe the selection methods likely to be used within two footprints: the WFD area, and the Roman Space Telescope's HLS. These two footprints

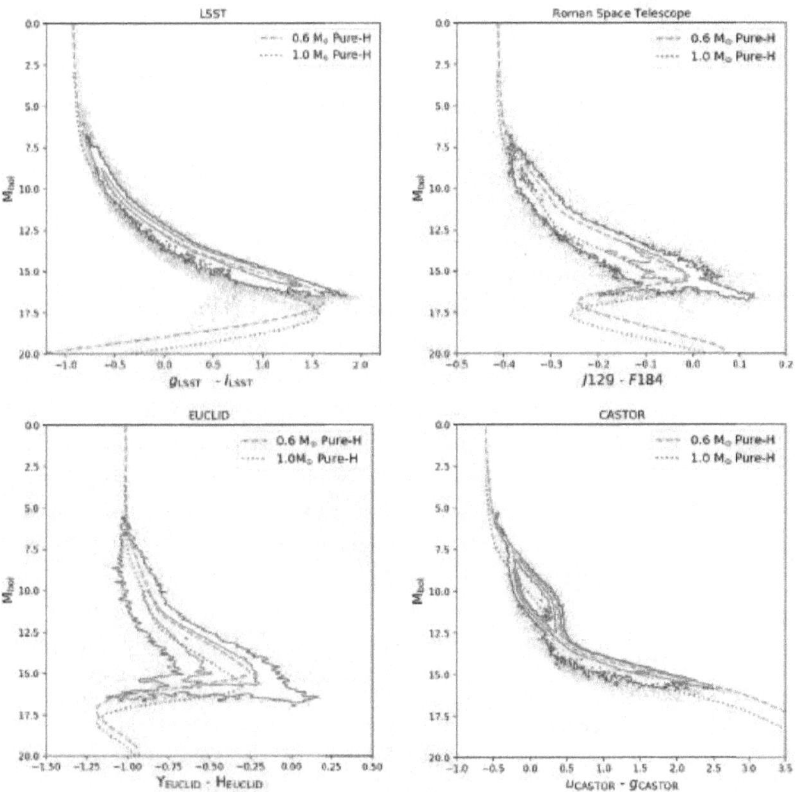

Figure 5.4 Colour-magnitude diagrams for the four surveys with model $0.6\,\mathrm{M}_\odot$ pure-H (solid cyan) and $1.0\,\mathrm{M}_\odot$ pure-H (dotted red) cooling curves over-plotted. The simulation includes the addition of photometric uncertainties.

will be the main focus of future white dwarf studies, a few of which we highlight in the following section.

5.5.1 The LSST WFD Survey Area

The science goals explored in §5.6 will depend on our ability to select samples from the large white dwarf populations presented in the previous section. Historically, white dwarfs have been predominantly selected by either relying on photometry combined with parallaxes or proper motions. Of course, the efficiency of these methods depends on the precision of the derived values. To explore the samples expected to have adequate parallax and/or proper motion accuracy, we used results from the LSST survey simulator, OPSIM (Connolly et al., 2014; Delgado & Reuter, 2016; Reuter et al., 2016), and applied them to our model.

The resulting distributions of parallax and proper motion precision can be seen in Figure 5.5, where the top panels show exponential fits to the results from OPSIM (black points). These functions are used to apply an uncertainty value to our model parallax and proper motions, and the bottom panels show the fraction of values, as a function of magnitude, above 5 and 10σ precision. The *Gaia*-based white dwarf catalogue of Gentile Fusillo et al. (2019) is plotted as a comparison, showing that the parallax precision for LSST will be similar to *Gaia*, although the proper motions will extend well beyond the *Gaia* magnitude limit. Below, we discuss the three "selection regimes" for LSST white dwarfs and present the resulting numbers in Table 5.2.

Regime 1: Selection on Parallax, Proper Motion, and Photometry

The most accurate method for selecting white dwarfs relies on using their parallaxes. Because they are intrinsically fainter than main-sequence stars at equal temperatures, white dwarfs occupy a well-defined region in the colour-magnitude diagram with minimal contamination from hot sub-dwarfs (see Gaia Collaboration et al., 2018a). This method was applied with great success in *Gaia*, which measured parallaxes for more than a billion stars over the whole sky.

Gentile Fusillo et al. (2019) used the *Gaia* DR2 catalogue to select white dwarf candidates based on their position in the *Gaia* colour-magnitude diagram and imposed a variety of *Gaia* quality flags to remove poor measurements. A comparison of this catalogue and the OPSIM LSST simulation suggests that the LSST catalogue will have similar precision to the *Gaia* mission, albeit over somewhat less than half the

Table 5.2 Number of White Dwarfs in Each Survey

Survey	Area deg^2	N_{WD}	N_{WD} $\Delta\pi > 10\sigma$	N_{WD} $\Delta\pi > 5\sigma$	N_{WD} $\Delta\mu > 10\sigma$	N_{WD} N_{WD} $\Delta\mu > 5\sigma$	N_{WD} HLS	N_{WD} $\Delta\pi > 5\sigma$ HLS	N_{WD} $\Delta\mu > 5\sigma$ HLS
LSST	18,000	154,849,052	140,090	286,387	3,500,461	6,819,463	1,240,950	22,473	360,229
Euclid	15,000	6,557,974					532,689		
Roman Space Telescope	2,213	735,401			703,072	726,714	735, 401	703,072	726,714
CASTOR	7,700	7,796,149					621,831		

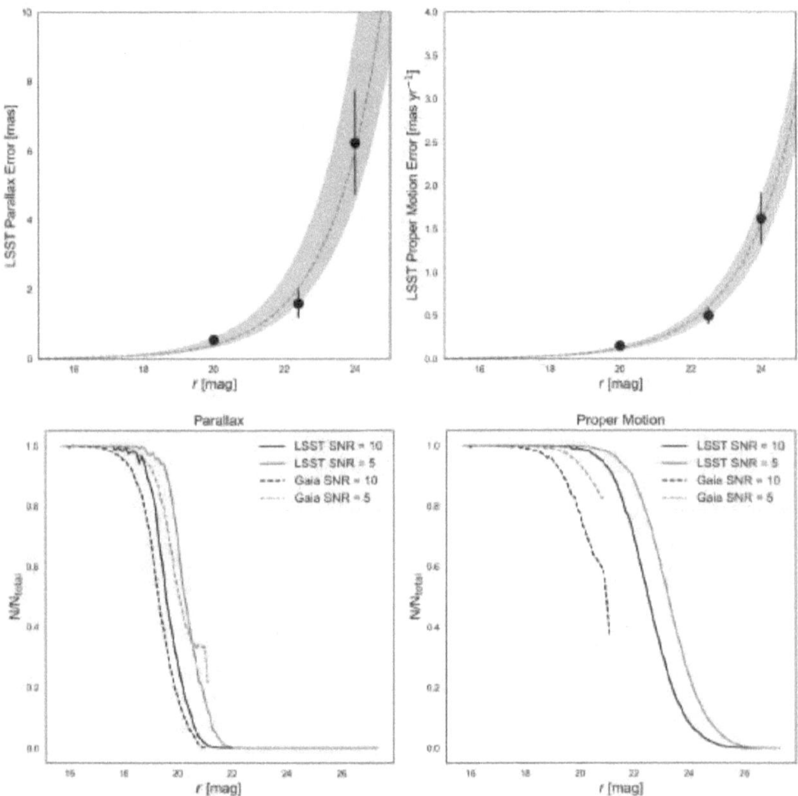

Figure 5.5 *Top:* The astrometric precision from the LSST OPSIM is fit using an exponential function (grey region). The same is done for the proper motion precision as a function of observed r-band magnitude. *Bottom:* The result of applying both of these prescriptions to the simulated dataset, showing the fraction of objects having greater than a 10σ (black) and 5σ (red) precision on the proper motion (bottom-right) or parallax (bottom-left). Also plotted are the results from the *Gaia* white dwarf sample presented in Gentile Fusillo et al. (2019) (dashed lines).

total area. However, this may be a lower limit as many of these white dwarfs are expected to be nearby. At faint magnitudes, this will result in a larger parallax measurement than the average point-source. Given that the precision of the parallax measurement is a function of distance, it is possible that LSST will measure accurate parallaxes for white dwarfs beyond what is suggested by OPSIM.

A stricter cut performed specifically on the parallax signal-to-noise ratio (SNR) was used by Kilic et al. (2019) to measure accurate distances to a set of white dwarfs having kinematics inconsistent with the Milky Way disk. The authors used a cut of 5σ on the parallax as this reduced the uncertainty on their calculated radius, which is derived using multi-band photometry combined with the parallax. The bottom left-hand panel of Figure 5.5 shows the 5σ error curve for the parallax compared to the *Gaia* dataset from Gentile Fusillo et al. (2019), showing roughly equal performance as a function of magnitude, with LSST sample extending a few tenths of a magnitude fainter. Thus, as with the *Gaia* dataset, the majority of objects having $r \lesssim 20.0$ mag will be able to make use of 5σ parallax measurements.

Regime 2: Selection on Proper Motion and Photometry

Below *Gaia*'s limiting magnitude of $r \sim 20$, LSST's full impact will be achieved by providing proper motion measurements down to its single-epoch, 5σ photometric limit of $r \sim 24.5$.

The lower right-hand panel of Figure 5.5 shows the fraction of objects, as a function of r-band magnitude, which will have greater than 5σ (solid red) and 10σ (solid black) measurements for their proper motion. We include the *Gaia* DR2 sample of Gentile Fusillo et al. (2019) for reference.

As with parallax measurements, the intrinsic faintness of white dwarfs means that, at equal temperature or colour, a white dwarf will exhibit a larger proper motion than stars belonging to other stellar classes. This property has made it possible for clean samples of white dwarfs within the SDSS footprint to be selected by Harris et al. (2006) and Munn et al. (2017) as well as in the SuperCOSMOS Sky Survey by Rowell & Hambly (2011). Such methods rely on the reduced proper motion of an object.

The studies of Rowell & Hambly (2011) and Munn et al. (2017) used proper motion cuts at 5σ and 3.5σ, respectively, to select their white dwarf samples. As Figure 5.5 shows, the approximate magnitude limit for which LSST will recover 50% of objects with proper motion measurements greater than 5σ is $r \sim 23.2$. This limit is 2-4 magnitudes deeper than the aforementioned studies. Selecting all objects with 5σ or better accuracy on their proper motion measurement results in nearly 7 million white dwarfs, two orders of magnitudes larger than previous studies.

Regime 3: Selection on Photometry

The majority of white dwarfs detected by LSST will be beyond the limits needed for accurate parallax and proper motion measurements. These objects, more than 140 million of them, will nevertheless have accurate six-band photometry down to the full depth of the survey. The primary obstacle with identifying white dwarfs at these faint magnitudes is that the white dwarf cooling sequence will overlap with many other stellar classes within a colour-colour diagram, particularly at cooler temperatures (see the top panel of Figure 1.4).

Photometric selection, therefore, focuses on hot white dwarfs within a colour-colour diagram. A selection of this manner was performed in Fantin et al. (2017) using data from the Next Generation Virgo Cluster Survey finding a contamination rate of 14% for white dwarfs with temperatures above 12,500 K. This contamination rate, however, was measured using SDSS spectroscopy which is incomplete and typically contains objects brighter than $r = 21.0$ mag. A much smaller contamination rate (nearly zero) was found by Bianchi et al. (2011) using GALEX UV photometry, which again was determined to the limit of the SDSS spectroscopic capabilities.

While we cannot extrapolate to the depth of the LSST, the rate of contamination will certainly be much higher when imaging nearly 7 magnitudes fainter. This regime will need to rely on the full spectral energy distribution (SED) of the objects to determine the probability of an object being a white dwarf. Several large spectroscopic surveys, such as WEAVE, DESI and 4MOST, will aid in the creation of large, diverse training sets for such endeavours.

Star-Galaxy Separation

One important caveat to these selection methods is the ability to separate Milky Way stars from background galaxies which become more distant and compact at fainter magnitudes. For example, a recent study by Slater et al. (2020) showed that the number of background galaxies is equal to the number of stars at $r \sim 20.5$ and increases to nearly 13:1 at $r=24.0$ and 25:1 at $r=25.0$. This means that accurate star galaxy separation will be important for Regime 2, but critically so for Regime 3.

Since background galaxies should have zero proper motion, the selection of white dwarfs should be possible up to the faint end of Regime 2 given the 5σ proper motion accuracy. As Slater et al. (2020) showed, LSST should be able to classify 80% of stars as being stars down to a limit of r-band magnitude of $23 - 23.5$ based solely on shape parameters derived from images. Combining this information with proper motion measurements should allow white dwarfs to be well distinguished from the background galaxies when accurate proper motions are available.

In Regime 3, however, shape measurements combined with the full 6-band photometry will need to be used to separate white dwarfs from background galaxies, and the efficacy of such an approach is beyond the scope of this work (for work on this topic, see e.g, Fadely et al., 2012; Soumagnac et al., 2015; Kim & Brunner, 2017; Bai et al., 2019; Slater et al., 2020).

5.5.2 The Roman Space Telescope High Latitude Survey

While LSST will soon uncover a huge number of white dwarfs, this sample will be made even more powerful by taking advantage of its overlap with other surveys. Space-based facilities like Euclid, the Roman Space Telescope, and CASTOR will dramatically improve star-galaxy separation over large portions of the WFD survey area since these facilities will have nearly $5\times$ better image quality. Moreover, these space-based surveys will provide photometric data at ultraviolet and NIR wavelengths, allowing for more complete characterization of white dwarf SEDs and the improved selection of double-degenerate and white dwarf/main-sequence binaries. Meanwhile, multi-epoch observations will make it possible to derive relative proper motions between the surveys.

To highlight the research opportunities that would be made possible by multi-wavelength and multi-epoch observations within a large survey region common to all facilities, we consider the Roman Space Telescope's HLS. This survey spans an area of 2,200 deg^2 in the Southern Sky, located far from the Galactic disk (b < -40, see Figure 5.1). While the goal of the HLS is to study dark energy by measuring the redshifts of more than 20 million galaxies and the shapes of more than 500 million galaxies, the deep multi-band infrared space-based imaging will provide a wealth of Galactic science cases as well (Spergel et al., 2015).

The Galactic science cases will be bolstered by the addition of kinematic information to the depths of the HLS. According to current plans, the Roman Space Telescope

HLS will be executed over 5 years which should allow for the measurement of proper motions. This topic has been explored by Sanderson et al. (2017) who estimated that the Roman Space Telescope will be able to produce relative proper motions with a precision $25\,\mu$as/yr to the depth of the survey ($J_{\mathrm{AB}} \simeq 26.9$ mag). This value is roughly four times better than proper motions achieved over a similar baseline *HST* ($100\,\mu$as/yr, see, e.g, Deason et al., 2013). However, if only a precision similar to *HST* is achieved, then it will still revolutionize the selection of white dwarfs. The ability to combine positions across surveys, the improved precision that may result from an extended *Gaia* mission, and optimized scheduling, could further improve the actual astrometric precision while allowing for an accurate transformation from relative to absolute proper motions (Sanderson et al., 2017).

One advantage that cool white dwarfs will have is that their intrinsic faintness means that they will be located much closer to the Sun than other objects observed at the limit of the survey. This means that these white dwarfs will experience a much larger proper motion, and as a result, the SNR of any measured proper motion will be higher. For example, a 3,500 K halo white dwarf at a distance of 750 pc present in our model will still experience a proper motion of 42 mas/yr, resulting in a total motion of 210 mas throughout the 5 year nominal survey. With a precision of 1.1 mas in position, this object will move nearly 200x this value. Similar objects ($M_{bol} \sim$15-17) experience a typical proper motion of 12 mas/yr in the thin disk, 22 mas/yr in the thick disk, and 50 mas/yr in the halo, moving a combined 60-250 mas throughout the survey — well above the positional precision of a single epoch measurement.

If we apply the astrometric precision from Sanderson et al. (2017) to our sample of HLS white dwarfs, we find that 99% of the white dwarfs in the HLS would have a proper motion detected at a 5σ level or better. This would allow a selection comparable to that of Regime 2 for the LSST to the full depth of the HLS ($J_{\mathrm{AB}} \sim$26.9), allowing many of the oldest and coolest white dwarfs to be detected with ease. We first present the results of the HLS simulation before exploring the implications of these results in the following section.

Figure 5.6 shows the results of our white dwarf simulations for the HLS, with the halo component highlighted in the right-hand column. The numbers within each sample are also included in Table 5.2. With each sample distributed over the same survey area, the difference in survey depths becomes apparent in the top panel of Figure 5.6.

The bolometric magnitude distributions reflect the passbands used in each survey

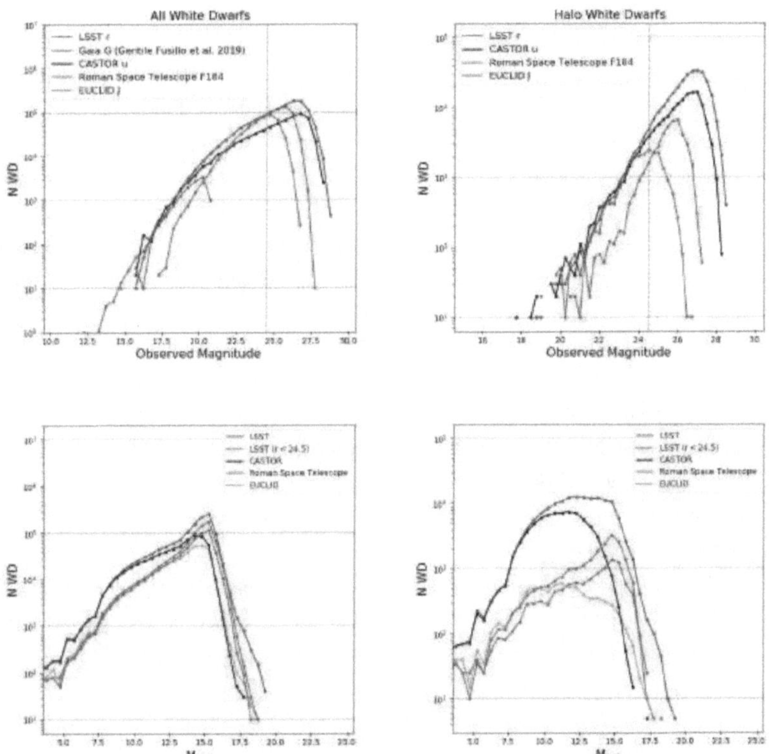

Figure 5.6 *Top:* The observed magnitude distributions for all white dwarfs (left) found within the Roman Space Telescope HLS, with the halo highlighted in the right column. These magnitudes include the effect of interstellar dust extinction as detailed in Section 5.3. *Bottom:* The bolometric magnitude distribution for each survey within the Roman Space Telescope HLS for all white dwarfs (left) and just the halo population (right).

since white dwarfs have SEDs similar to blackbodies (see Figure 5.1). The NIR surveys, the Roman Space Telescope and Euclid, will uncover a large number of cool white dwarfs, whereas the optical/UV surveys, CASTOR and LSST, will uncover the hotter, more recently formed white dwarfs. The Roman Space Telescope alone will uncover a few thousand halo white dwarfs beyond the turn-off in the white dwarf luminosity function, making it the go-to survey within the HLS footprint when looking to determine the age and star formation of the Galactic halo. The ability to cross-match these surveys over such a large area means that white dwarfs of all temperatures will be observed, providing a more complete census of the true Galactic population.

In Figure 5.7, we show the distance distribution of white dwarfs detected within the HLS field. Given that the Roman Space Telescope and Euclid surveys mainly target the intrinsically faint, cool white dwarfs, the volume of space which they survey is relatively small compared to CASTOR and LSST. CASTOR, in particular, will detect young white dwarfs up to distances of 30 kpc, although most will lie within a few kpc of the Sun, still far beyond current capabilities.

Our results show that within the HLS footprint, the selection of white dwarfs will rely heavily on the proper motions derived by the Roman Space Telescope, as the combination of precision and depth will surpass that of the LSST. Furthermore, the superior image quality will greatly aid in star-galaxy separation at the limits of the survey, allowing for a clean sample selection. With the added depth it is also possible to select cool white dwarfs at a greater distance than within the WFD survey, although, given the intrinsic faintness of these objects, even the distances probed by the Roman Space Telescope will be mainly within 1 kpc of the Sun.

5.6 White Dwarf Science Cases

With an understanding of the different catalogues and their limitations, this section explores some key white dwarf science cases. We emphasize that these are by no means comprehensive. Various science cases make use of different samples as defined in §5.5.1 and 5.5.2 with the absolute numbers within both the WFD and the HLS footprints presented in Table 5.2.

5.6.1 The White Dwarf Luminosity Function

The white dwarf luminosity function (WDLF) has been used in the past to determine ages and star formation histories for a variety of stellar populations, from the solar

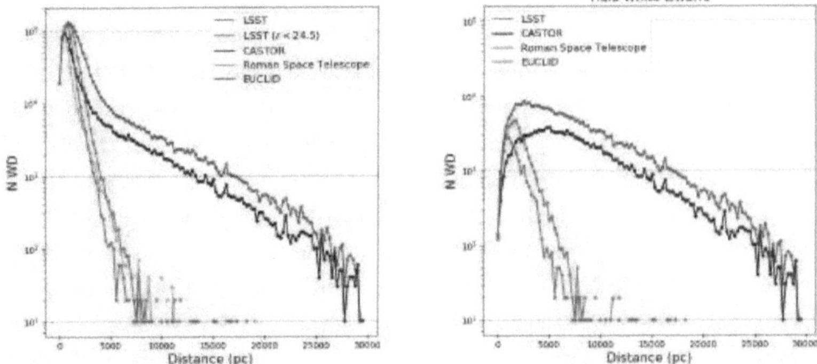

Figure 5.7 *Left:* Distance distribution of all white dwarfs within the HLS field seen by the various surveys. The UV surveys will observe hot white dwarfs over a much larger volume owing to their higher intrinsic brightness. *Right:* Halo white dwarf distance distribution in the HLS field showing a similar result.

neighbourhood to more distant star clusters(see, e.g, Richer et al., 2006; Bedin et al., 2008, 2009; Rowell, 2013; Torres et al., 2015; Salaris & Bedin, 2018). This is a consequence of a decrease in the cooling rate of white dwarfs at low temperatures combined with the finite age of the Universe. The result is that white dwarfs become more numerous at low temperatures before dropping off below this regime as objects will not have had enough time to cool further. This manifests itself as a turn-off in the WDLF.

Following the large increase in the number of known white dwarfs from surveys such as the SDSS, the WDLF has been determined with greater precision as the fainter magnitude limits increased the number of white dwarfs found beyond the observed turn-off. For example, Harris et al. (2006) found four stars beyond the turn-off at $M_{bol} = 15.25$, and Munn et al. (2017) found on the order of a hundred. This resulted in Kilic et al. (2017) providing the most precise age determination of the Milky Way's components to date using white dwarfs. Their results imply that the thin disk formed 7.4-8.2 Gyr ago, the thick disk at 9.5-9.9 Gyr, and the halo at $12.5^{+1.4}_{-3.4}$, although a turn-off in the halo luminosity function has yet to be observed. The addition of proper motion data from *Gaia* DR2, in combination with photometric data from CFIS and PS1, was used by Fantin et al. (2019) to determine the onset of star formation in the disk to be (11.3 ± 0.5) Gyr before reaching a maximum star-formation rate at (9.8 ± 0.3) Gyr, followed by a slight decline towards the present day. Given that the turn-off is the fundamental part of age determinations, finding objects beyond this turn-off is paramount and will require deeper surveys with accurate selection methods such as the WFD or HLS samples.

In Figure 5.8 we show the predicted WDLF for the LSST WFD survey using objects with 5σ proper motion measurements and compare it to results in the literature. Our WDLF was calculated using the $1/V_{max}$ method, which has been used for previous studies of observational WDLFs. This method, introduced by Schmidt (1968), aims to calculate the number density by summing the maximum volume each observed object could occupy and still be observed based on the survey parameters. Here, V_{max} is defined as

$$V_{max} = \beta \int_{r_{min}}^{r_{max}} \frac{\rho}{\rho_\odot} R^2 dR, \qquad (5.1)$$

where β is the solid angle subtended by the survey, $\frac{\rho}{\rho_\odot}$ is the density distribution of the component and

$$r_{min} = 10^{0.2(m_{min}-M)} \qquad (5.2)$$

$$r_{max} = 10^{0.2(m_{\max} - M)} \tag{5.3}$$

are the minimum and maximum distances for which the object would be observed given the bright and faint magnitude limits of the survey.

We use the density distributions as presented in §5.3 while the solid angle is calculated using the survey parameters. The number density is then the sum of the maximum volumes for each object within each bolometric magnitude bin,

$$\phi = \sum_{i=1}^{N} \frac{1}{V_{\mathrm{max},i}} \tag{5.4}$$

while the uncertainty follows Poissonian statistics for each bin,

$$\sigma_\phi^2 = \sum_{i=1}^{N} \frac{1}{V_{\mathrm{max},i}^2}. \tag{5.5}$$

Our results show a disk luminosity function that increases from $M_{\mathrm{bol}} = +4$ to $+15.25$ mag before dropping off sharply. This drop-off is a consequence of the finite age of the universe combined with the mass distribution of white dwarfs. Since high-mass C/O white dwarfs cool faster than their low-mass counterparts, we see a significant number of high-mass white dwarfs beyond this turn-off. These objects, with masses around 0.8–$1.0\,M_\odot$, are the remnants of intermediate-mass stars formed between 8 and 12 Gyr. For the LSST, we find there will be nearly 100,000 of these objects in the 5σ proper motion sample.

Our halo luminosity function is presented in the right-hand panel of Figure 5.8. The results show that the increased depth of the LSST will detect the turn-off in the halo luminosity function for the first time, with more than 40,000 objects below $M_{\mathrm{bol}} = 15.5$.

5.6.2 The Initial-to-Final Mass Relation

As the end stage of more than 97% of stars, white dwarfs are important tests of stellar evolution models. For a given initial (main-sequence) mass and age, their present-day masses contain information on the amount of mass lost during the progenitor's post-main-sequence evolution.

The relation between the initial mass is given by the IFMR. While the IFMR does have theoretical prescriptions (e.g, Marigo & Girardi, 2007; Choi et al., 2016),

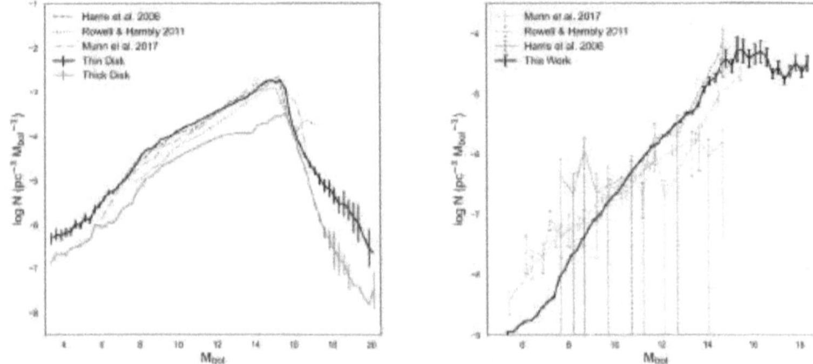

Figure 5.8 *Left:* The luminosity function for thin and thick disk white dwarfs found within the LSST 5σ proper motion sample. Also plotted are results from Harris et al. (2006), Rowell & Hambly (2011), and Munn et al. (2017). *Right:* The halo white dwarf luminosity function with the same cut in proper motion precision.

it is usually measured semi-empirically. While these methods can include using wide double degenerate binaries (Andrews et al., 2015) or the color-magnitude diagram (El-Badry et al., 2018) we focus on the star cluster method. Star clusters can be used as their known age can be translated into an initial mass with an appropriate stellar isochrone. Thus, by measuring the masses of newly formed white dwarfs in a cluster, one can relate the initial stellar masses to the final white dwarf masses. By combining measurements over a wide range of ages (and hence initial masses), an empirical relation can be established between the two quantities (see e.g, Kalirai et al., 2008; Cummings et al., 2018).

The combination of LSST and CASTOR will likely provide the best data for constraining the precise form of the IFMR. These surveys are sensitive to the hot, recently formed, white dwarfs needed to use this technique, though ultimately, this relies on follow-up spectroscopy to measure masses. That is to say, these surveys will provide the imaging needed to identify hot white dwarfs in distant clusters (up to 20 kpc as shown in Figure 5.7) which can be targeted for spectroscopy with large optical telescopes.

The left-hand panel of Figure 5.9 shows the observed g-band magnitude for a young white dwarf ($T_{\mathrm{eff}} = 25,000\,\mathrm{K}$) as a function of distance for four different masses. For reference, the halo is currently producing white dwarfs with masses close to $0.52\,M_{\odot}$, while younger open clusters are producing white dwarfs between $0.6\,M_{\odot}$ and $1.0\,M_{\odot}$ depending on their age. As Figure 5.9 shows, many of the hot white dwarfs in star clusters closer than $\sim 7.5\,\mathrm{kpc}$ will be detected by LSST.

We have used a modified version of our model to simulate the stellar population in the globular cluster Messier 10 (M 10) to highlight the potential sample of high-temperature white dwarfs which could be used for such a study. The resulting simulation can be seen in the middle panel of Figure 5.9. We have set the mass of the cluster to be $2.25 \times 10^{5}\,M_{\odot}$ with structural parameters as described in Gnedin et al. (1999). The simulation predicts nearly 75,000 white dwarfs present within M 10, of which 7,500 are bright enough to be observed by the LSST 10-year survey (blue points). In this sample, we find that 63 objects have $T_{\mathrm{eff}} > 25,000$ K (red points), with a few lying outside the core region.

While crowding will play a role in the final number of objects discovered, our results suggest that there will be, at minimum, a handful of hot white dwarfs observed in the outskirts of the cluster. Furthermore, since these objects are forming from low-mass stars, mass segregation may increase the number of progenitors in

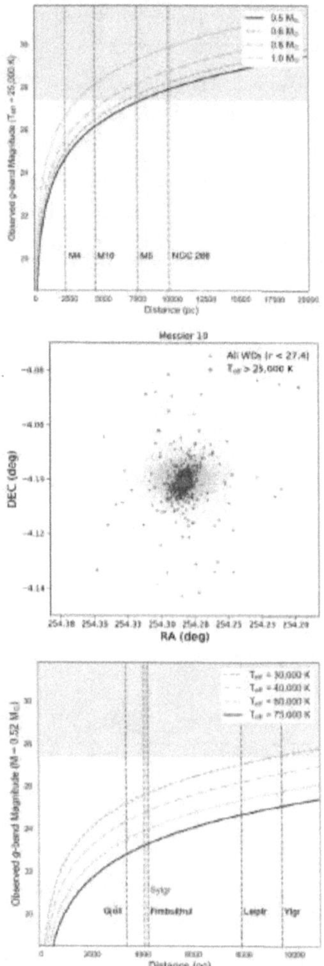

Figure 5.9 *Top:* The observed g-band magnitude of a young white dwarf as a function of distance for four masses. The grey shaded region represents magnitudes beyond the 5σ g-band magnitude limit of the LSST 10-year survey. Also marked are four globular clusters within the LSST footprint, of which one (NGC 288) also lies within the Roman Space Telescope HLS survey. *Middle:* Simulation showing the white dwarf population in Messier 10, with objects hotter than 25,000 K highlighted in red. *Bottom:* The magnitude of young, hot, white dwarfs as a function of distance. Also marked are four nearby streams located within the LSST WFD footprint discovered by Ibata et al. (2019).

the outskirts of the cluster leaving a larger number of recently formed white dwarfs (Spitzer, 1987; Parada et al., 2016a). These photometric data, combined with upcoming large spectroscopic surveys like the proposed Maunakea Spectroscopic Explorer (Hill et al., 2018; The MSE Science Team et al., 2019) or the next generation Extremely Large Telescopes, will allow for the most accurate semi-empirical determination of the IFMR.

5.6.3 Stellar Streams

Recent studies have shown that the inner halo is crisscrossed with stellar streams, many of which are the tidal debris from halo globular clusters (Malhan et al., 2018; Ibata et al., 2018, 2019). These streams were discovered using *Gaia* DR2, relying mainly on the measured proper motions and photometry. LSST will provide equivalent information to much fainter magnitudes, suggesting that it will likely increase the number of known stellar streams.

With the addition of proper motions at fainter magnitudes, more stars within the known streams will be identifiable. To determine whether any white dwarfs will be discovered within these streams, we select five nearby streams from Ibata et al. (2019) and simulate the magnitudes of a hot white dwarf within each stream. These streams — Gjöll, Sylgr, Fimbulthul, Leiptr and Ylgr — all lie within the LSST WFD footprint. Given that these streams are metal-poor, we assume that the stellar population is composed of stars similar to those in the oldest globular clusters (12 Gyr) so their hot, young, white dwarfs will be forming with masses of $\sim$0.52 $M_\odot$. We show the resulting magnitudes for these objects as a function of distance in the right-hand panel of Figure 5.9.

This exercise shows that LSST will provide g-band photometric measurements for the hot, young, white dwarfs ($T_{\mathrm{eff}} > 30{,}000\,\mathrm{K}$) in all five streams, with the hottest objects ($T_{\mathrm{eff}} > 50{,}000\,\mathrm{K}$) in the nearest streams (Gjöll, Sylgr and Fimbulthul) likely having reliable proper motions.

The ability to reliably assign membership to these streams will rely on the measured properties of the stream, including the distance and proper motion, that will be determined from brighter more numerous stars. The addition of these white dwarfs could provide information on the age if accurate masses can be measured. Because these streams can be closer to the Sun than many halo globular clusters, they could provide additional information on the low-mass end of the IFMR — provided the stream age can be determined independently.

5.6.4 Pulsating White Dwarfs

One of the key science cases for time-domain surveys like the LSST is the study of luminosity fluctuations in stars. The first pulsating white dwarf, HL Tau 76, was discovered by Landolt (1968). Subsequently, it was discovered that these objects had similar temperatures, leading to improved selection methods (Fontaine et al., 1980). The discovery of this new class of white dwarf named after the prototype, ZZ Ceti, was important as it allowed for the study of the mass of the stratified interior layers of the white dwarf, while also allowing for an independent determination of the mass and atmospheric parameters (see, e.g, Kepler et al., 1995; Pech & Vauclair, 2006; Pech et al., 2006; Castanheira & Kepler, 2009; Althaus et al., 2010; Romero et al., 2012; Calcaferro et al., 2018).

While there are different types of variable white dwarfs, the vast majority of the currently known white dwarf variables belong to the ZZ Ceti class. These objects have temperatures between 11,000 and 13,000 K with pure-Hydrogen atmospheres, representing the coolest class of pulsating white dwarfs (Van Grootel et al., 2012). White dwarfs with pure-Helium atmospheres also exhibit pulsations (and are classified as DBV), as do extremely low-mass white dwarfs (ELMV and pre-ELMV). The remaining classes include much hotter white dwarfs, many of which are still surrounded by their planetary nebula (Córsico, 2018). An excellent comprehensive review of these classes and the current observational state of pulsating white dwarfs can be found in Córsico et al. (2019). Here, we focus on the ZZ Ceti regime while noting that LSST will also uncover a large number of the other classes. In particular, LSST should be able to accurately characterize the instability strips for the other classes of white dwarf pulsators as their rarity requires this type of wide-field, multi-epoch survey.

ZZ Ceti stars experience pulsations as a result of the increased opacity in their outer atmosphere due to the recombination of hydrogen. This typically occurs between 11,000 and 13,000 K, with this temperature range being a function of mass. Fontaine et al. (1982) suggested that all DA white dwarfs will experience these pulsations (on the order of 100-1400 s) as they cool and move across this instability strip, and observational evidence has since confirmed this hypothesis (Fontaine & Brassard, 2008).

Therefore, to study the number of potential ZZ Ceti stars in the LSST, we use the observational boundaries from Van Grootel et al. (2012) and apply them to our simulated sample of DA white dwarfs with $>5\sigma$ proper motion objects. The boundary and results can be seen in Figure 5.10, where the left-hand panel shows the selection region, and the right-hand panel shows their location within a colour-colour diagram. We note that our model includes many massive objects above the specified mass regime of Van Grootel et al. (2012). Recent observations have uncovered a few examples of these massive pulsators ($1.20 \pm 0.03\,\mathrm{M_\odot}$, see Hermes et al., 2013), suggesting that the selection regime could be extended to the upper mass limit.

The total number of objects lying within this region is $\sim$200,000 in our 5σ proper motion sample, with nearly 8,000 being brighter than the faintest ZZ Ceti discovered to date (g = 19.5, see Table 5 in Córsico et al., 2019).

Finally, we note that although we used the proper motion selected sample, ultimately the ability to study the pulsations will rely on the photometry and the ability to follow-up the observations at a higher cadence to more accurately determine periods. These objects will also benefit from follow-up spectroscopy to accurately determine their temperature and mass to define the boundaries of their respective instability strip.

5.6.5 Metal Polluted White Dwarfs

Our focus to this point has been on white dwarfs with pure-Hydrogen or pure-Helium atmospheres as they are the most abundant populations. The atmospheres of white dwarfs, however, can also be polluted by metals. If the metal lines are the strongest spectral features the object is designated as spectral type DZ, although metal pollution can occur in other spectral types as well. For example, DA white dwarfs with metal lines present are designated as DAZ. The ability to detect such objects are of interest as the occurrence rates can reveal information regarding their formation mechanisms.

Metal-polluted white dwarfs can form as a result of episodic accretion from planetary material since any metals present in the photosphere at the time of formation would rapidly sink to the core (Paquette et al., 1986). This means that any metals present in the white dwarf atmosphere can be used to constrain the bulk composition of the accreted bodies (Zuckerman et al., 2007), which has revealed a large amount of diversity in the planetary systems surrounding white dwarfs (Hollands et al., 2018a).

Such samples have typically relied on no more than a few hundred objects, how-

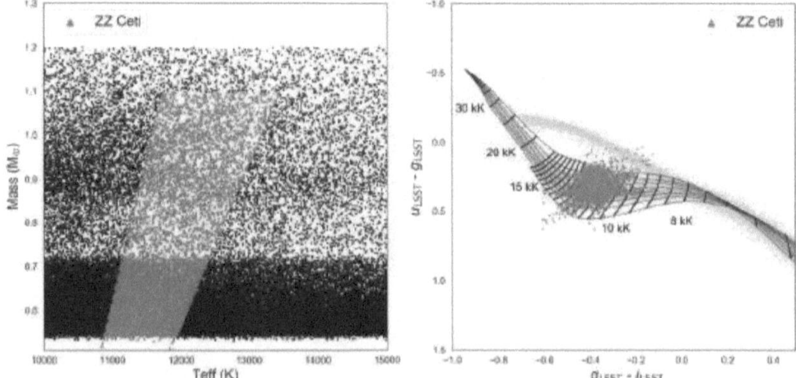

Figure 5.10 *Left:* $\mathrm{T_{eff}}$–mass plot for white dwarfs in our LSST 5σ proper motion sample (black points) highlighting the location of the ZZ Ceti instability strip defined by Van Grootel et al. (2012) (red points). *Right:* The location of these objects in a colour-colour diagram, with solid black lines representing lines of constant $\mathrm{T_{eff}}$ and faded lines representing constant mass. Temperature values decrease with increasing $(g - i)$ colour as indicated.

ever, the incidence of metal pollution in white dwarfs has been found to be substantial. An example of this is the study of DA white dwarfs with cooling ages between 20 and 200 yr (T_{eff} = 22,000 - 27,000 K) performed by Koester et al. (2014) using *HST* UV spectroscopy. Their results suggest that at least 27% of these objects show evidence of recent accretion, and the value may be closer to 50%. LSST will inevitably observe large numbers of these objects and will be critical in identifying candidates for follow-up spectroscopy as many of the metal lines exhibited by these objects (Ca, Mg, Na) have strong absorption lines in the near-ultraviolet which results in decreased flux in the u-band. Indeed, the number of objects in our WFD 5σ proper motion samples within the temperature range studied by Koester et al. (2014) is nearly 550,000.

Koester et al. (2011) and Hollands et al. (2017) used the SDSS photometry, in particular the u-band, to identify cool metal-polluted white dwarfs in the SDSS. These objects separate from other white dwarf types below $\sim$8,000 K as the absorption from Ca I begins to dominate over the Balmer series within the u-band. To show this behavior, we have simulated metal-polluted white dwarfs with three different calcium abundances in the LSST and CASTOR photometric bands as both surveys will contain a u-band. These models are adopted from those presented in Coutu et al. (2019). The three different abundances represent the regimes between the most metal-polluted white dwarfs ([Ca/He] = -7) to those with little pollution ([Ca/He] = -12). The resulting colour-colour diagram for LSST is presented in Figure 5.11, with the selection regime of Koester et al. (2011) and Hollands et al. (2017) highlighted as the dashed box. As with the SDSS, LSST will allow for a separation of white dwarfs in colour-colour diagrams as it has very similar passbands; however, the selection will be more accurate given the increased depth of the LSST and the ability to distinguish white dwarfs from other point-sources via their proper motions.

We have also explored the possibility of selecting metal-polluted white dwarfs over a wider temperature range in the right-hand panel of Figure 5.11. Here, we have cross-matched LSST with CASTOR in an attempt to isolate the warmer objects. Focusing on the [Ca/He] = -7 curve shows that the addition of a UV filter and a broad wavelength coverage can indeed separate hot metal-polluted objects with a large amount of calcium absorption at 12,000 K. This suggests that a combination of colours can be used to identify a larger sample of metal-polluted white dwarfs for follow-up observations with new and upcoming spectroscopic instruments, the results of which will allow for the study of the rate of incidence and physical processes at work in such systems.

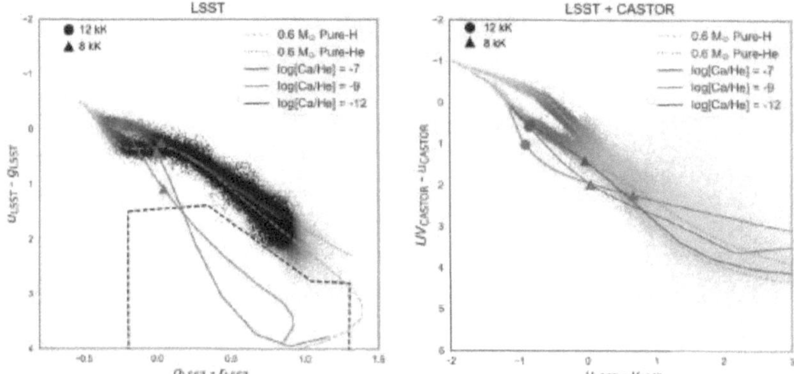

Figure 5.11 *Left:* LSST colour-colour diagram showing the cooling tracks for pure-H, pure-He, and pure-He contaminated with Ca for which we use as a representation for a DZ white dwarf. The black dashed box represents the colour region used by Koester et al. (2011) and Hollands et al. (2017) to select cool DZ white dwarfs. The circles on each plot represent $T_{eff} = 12{,}000\,$K, while the triangles represent $T_{eff} = 8{,}000\,$K. *Right:* By combining LSST with other CASTOR it is possible to also separate higher T_{eff} DZ white dwarfs, suggesting that a combination of colours can be used to select DZ white dwarfs over a range of temperatures.

5.6.6 Debris Disks and Sub-stellar Companions

The metals present in white dwarfs are thought to originate from the accretion of either debris disks or remnant planetesimals (see e.g, Kilic et al., 2006; Farihi et al., 2010; Koester et al., 2014). The debris disks are warm, with temperatures ranging from a few hundred to a few thousand Kelvin and occur around $1-5\%$ of white dwarf (Barber et al., 2012), although the fraction of systems showing evidence for recent accretion may be as high as 50% (Koester et al., 2014).

These debris disks are typically discovered by comparing model SEDs of the white dwarf using properties determined through optical spectroscopy to the observed infrared magnitudes, where any excess infrared emission is attributed to the debris disk. We have highlighted this point in Figure 5.12, where we have shown the SED of a metal-polluted white dwarf with a surface temperature of 12,500 K and surrounded it by a debris disk of 1,200 K. This disk is modelled after the G29-38 (Zuckerman & Becklin, 1987; Reach et al., 2005) disk presented in Jura (2003) (see equation 3). This disk is modelled as a face-on, flat, opaque ring surrounding the host star with an inner temperature of 1,200 K at 0.14 $R_\odot$. We set the distance to be 100 pc from the Sun.

The debris disk is completely invisible over the optical wavelengths covered by the LSST, although the excess emission from the debris disk becomes apparent in the F184 bands from the Roman Space Telescope, with a difference in observed magnitude of 0.27 AB mag. This difference should be detectable to a limit of $\sim$25th magnitude in the F184 band. This shows that the combination of optical data from the LSST, combined with the NIR photometry from the Roman Space Telescope or Euclid, will be ideal for detecting such objects.

To estimate the number of white dwarfs with debris disks that might be present within the HLS, we apply the selection method from Wilson et al. (2019), $T_{\mathrm{eff}} = 14,000 - 31,000$, and apply their resulting rate of debris disks which show an infrared excess, $1.5^{+1.5}_{-0.5}\%$, to our sample of LSST 5σ proper motions combined with the Roman Space Telescope sample. Our results suggest that this method could detect on the order of 355 white dwarfs with debris disks in the HLS footprint.

We note, however, a few caveats that come with this number. First, the disk modelled above is a more extreme example of debris disks surrounding white dwarfs as its inclination is thought to be close to zero. Rocchetto et al. (2015) showed that such bright, face-on disks comprise less than 10% of all known white dwarf debris

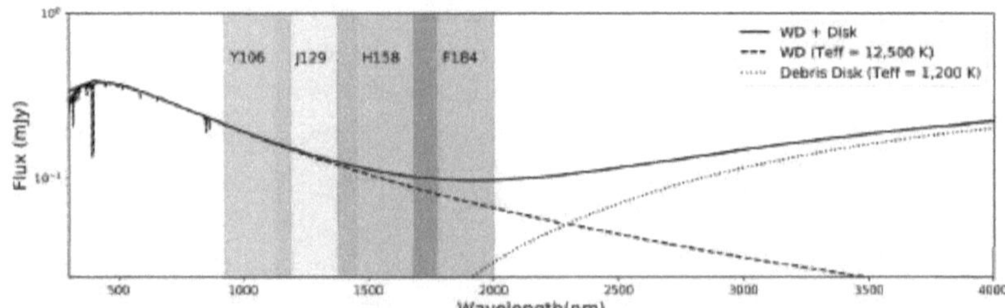

Figure 5.12 SED of a metal polluted white dwarf (12,500 K) with a warm debris disk (T = 1,200 K) with the Roman Space Telescope filters highlighted. The presence of a debris disk will result in an excess of observed flux in the infrared, and will allow for the detection of such disks.

disks, while most have luminosities that are less than 20% of such a model. The remaining disks, as well as those with lower temperatures and/or at larger distances, will be more difficult to detect. As an example, changing the inclination of the G29-38 disk from face-on to an inclination of 75° decreases the flux excess in the F184 band to 0.07 AB mag, and decreasing the temperature, size, or opacity will reduce this value even further. Applying the results from Rocchetto et al. (2015) still suggests that a few dozen bright, dusty, debris disks surrounding white dwarfs with an excess in the F184-band will be detectable in the HLS.

Another caveat is that debris disks have been found in cooler white dwarfs (Kilic et al., 2006), and as such a larger number of white dwarfs with debris disks may await discovery. Ultimately, the combination of LSST and Euclid and/or the Roman Space Telescope will help in determining the fraction of white dwarfs with 2μm emission over a wide range of temperatures.

5.7 Summary

In this chapter, we have used a new model to simulate the white dwarf populations that will be detected in three pending and one proposed wide-field surveys (LSST, Euclid, the Roman Space Telescope and CASTOR) whose depth, wide fields and broad wavelength coverage across the UV/optical/NIR region will open new avenues in white dwarf research. Our key findings are as follows:

- We predict that LSST, at its final 10-year depth, will detect more than 150 million white dwarfs in its WFD survey, although identifying many of these white dwarfs could be a challenge, depending on the availability of parallaxes and/or proper motions, the precision of those measurements, and availability of supplemental data from other facilities.

- We identify three distinct magnitude regimes in which different parameters measured by LSST can be used to optimally select white dwarfs. At $r \lesssim 20.0$, we predict $\sim 300{,}000$ white dwarfs can be identified using parallaxes measured to a precision of 5σ or better. In the range $20.0 \lesssim r \lesssim 23.5$, we expect ~ 7 million white dwarfs can be identified using proper motions measured to a precision of 5σ or better. We anticipate the impact of LSST for white dwarf research will be highest in this regime owing to the tremendous gain in sample size and the generally low contamination rate. Fainter than $r \sim 23.5$, white dwarfs will need to be selected from photometry alone. This will be particularly difficult as contamination from other stellar contaminants and inter-loping background galaxies becomes substantial at faint magnitudes. The expanded wavelength coverage and excellent image quality provided by the space-based facilities will be especially important for identifying clean white dwarf samples at these faint magnitudes.

- Proper motions will continue to be a key tool for identifying white dwarfs within upcoming surveys. While LSST, and potentially the Roman Space Telescope observations acquired for its High Latitude Survey, will yield accurate proper motion measurements, the cadence adopted by other facilities or programs will be important for maximizing white dwarf yields.

- The Roman Space Telescope High Latitude Survey represents the largest over-lapping region between the four surveys. This region will benefit from excellent star-galaxy classification and simultaneous UV, optical and NIR photometry. Several white dwarf science cases, including from initial-to-final mass relation and the frequency of white dwarf debris disks and sub-stellar companions — will be enabled by the multi-wavelength observations in this region.

- With simulations, we illustrate how the white dwarf luminosity function — for both the disk *and* the halo — will be well constrained beyond the turn-off point, leaving only systematic uncertainties for all but the most extreme ends.

- UV observations will be particularly useful for identifying the hottest and youngest white dwarfs in star clusters out to 20 kpc. Simulations for the globular cluster Messier 10 show that a handful of young, hot, white dwarfs will be observed with the LSST. These objects will need spectroscopic follow-up with upcoming large telescopes to determine an accurate mass to derive an empirical initial-to-final mass relation.

- LSST will observe more than 200,000 ZZ Cetis, providing ~ 800 independent flux measurements over the 10-year survey lifetime; we estimate that more than 8000 of these objects will be brighter than $g = 19.5$. In addition, the survey will help constrain the instability strip boundaries for other types of pulsating white dwarfs.

- LSST will also observe white dwarfs with polluted atmospheres. We show that a combination of the UV and optical surveys will be useful for identifying white dwarfs with metal-polluted atmospheres over a wider range of temperatures when compared to data solely acquired as part of the LSST.

- NIR photometry is critical for the discovery of debris disks surrounding white dwarfs. Using current estimates for the fraction of white dwarfs having associated debris disks, we estimate that combined LSST and Roman Space Telescope observations will uncover dozens of white dwarfs with debris disks.

This study highlights only a few of the many exciting science cases involving white dwarfs during the coming decade, including the white dwarf luminosity function of the Milky Way's components, the initial-to-final mass relation, and the variable and pulsating white dwarfs, to name a few. LSST and other surveys will provide a multitude of science cases beyond what has been presented here, and examples in the literature include the discovery of companion planets (see e.g, Lund et al., 2018; Cortés & Kipping, 2019) and the study of white dwarfs through gravitational waves Korol et al. (2019). The sheer number of objects to be revealed in these surveys will inevitably lead to unforeseen discoveries, pointing the way to a decade of exciting white dwarf research.

Chapter 6

Conclusions

This chapter highlights the contributions this book has made to the field of white dwarfs and their use in studying the Milky Way, before looking ahead to the future of the field.

6.1 White Dwarfs as Tracers of Galactic Evolution

6.1.1 White Dwarf Population Synthesis Model

Recent wide-field surveys, like the Sloan Digital Sky Survey, have revolutionized the ability to use stellar populations as tracers of Galactic evolution by comparing observations to stellar population synthesis codes. While these studies can also be extended to post-main sequence stars, such as white dwarfs, one needs to model the complete evolution of a star in order to properly compare their resulting distribution to any input parameters. Since previous stellar population synthesis codes either added white dwarfs based on a local density (Besançon, Robin et al., 2003), or could not accurately reproduce observed distributions (TRILEGAL, Girardi et al., 2005) as shown in Fantin et al. (2017) I developed a fully self-consistent white dwarf population synthesis code and presented it in Chapter 2.

The model takes ten functional inputs: the star formation history, the density distribution, the initial-mass function, the initial-to-final mass relation, the stellar lifetimes, the metallicity, the white dwarf cooling tracks, the DA:DB fraction, and the Milky Way stellar velocity distributions and outputs a catalogue of white dwarfs. The model can also apply any observational constraints, such as magnitude limits and survey areas to the resulting outputs. This allows a direct comparison between selected model inputs and the resulting observations.

6.1.2 The Milky Way Star Formation History

The first application of this model was to determine the star formation history of the Milky Way by comparing the observed distribution of white dwarfs in the Canada France Imaging Survey (CFIS). This work was published in Fantin et al. (2019) and is presented in Chapter 3.

The observational white dwarf catalogue used in this work was obtained by matching the u-band data from CFIS with *grizy* data from the Pan-STARRS1 DR1 (PS1) catalogue and proper motions from *Gaia* DR2. This allowed two observational diagrams to be made, a colour-colour diagram and a reduced proper motion diagram, which is a combination of photometry and proper motions. These two diagrams served as the observational data for which the model would be compared to.

The model was modified to compute synthetic white dwarf photometry in the CFIS and PS1 bands. The CFIS survey area and *Gaia* photometric limits were also added to the model framework to account for these observational effects. The star formation history for each Galactic component was parameterized as a skewed Gaussian to maximize the number of degrees of freedom being fit by the model. The results showed that the Milky Way began by producing halo white dwarfs at a modest, albeit uncertain, rate before a large burst of star formation in the thick disk began at (11.3 ± 0.5) Gyr. This burst slowly declined as the thin disk began forming stars at (8.4 ± 0.3) Gyr until the present day at a rate of $(3.5 \pm 0.3)\,\mathrm{M_\odot yr^{-1}}$. Studying the residuals, however, revealed that the star formation history of the thin disk likely deviated from a single uni-modal function with an increase by as much as 50% at (3.3 ± 1.8) Gyr and a decrease of 30% at (5.8 ± 1.1) Gyr.

The model also determined both the local space densities of each component as well as their relative contributions. The white dwarf space densities were found to be $(4.8 \pm 0.4) \times 10^{-3}\,\mathrm{pc^{-3}}$ for the thin disk, $(1.0 \pm 0.2) \times 10^{-3}\,\mathrm{pc^{-3}}$ for the thick disk, and $(6.3 \pm 2.4) \times 10^{-6}\,\mathrm{pc^{-3}}$ for the halo, resulting in an $(83 \pm 5)\,\%$, $(17 \pm 3)\,\%$, and $(0.11 \pm 0.05)\,\%$ contribution from the thin disk, thick disk, and halo respectively.

6.1.3 Halo White Dwarfs

Another conclusion from Chapter 3 was that the star formation of the Galactic halo seen through its white dwarfs remained rather uncertain due to the small sample size relative to the disk. This motivated the analysis presented in Chapter 4, where I acquired follow-up spectroscopy of a sample of halo white dwarfs to measure their masses, which can be converted into an age.

The results suggest a wider variation in ages than expected from a single, $\sim$12 Gyr burst of star formation which is typically assumed. However, the fitting process likely suffered from a systematic bias in which the temperature-surface gravity degeneracy could not be broken in low signal-to-noise (SNR) data. This resulted in a likely overestimation of the mass. This result is presented in Figure 4.15 and shows a decline in mass as the signal-to-noise at the higher-order Balmer lines increases. Fitting sub-exposures of our low mass, high SNR objects did not show a decline in mass as the SNR increased, however, many of these sub-exposures were already above the minimum SNR determined by Kepler et al. (2006) that allows the temperature-surface gravity degeneracy to be broken.

This led us to split our sample based on their SNR. The masses for objects which has an SNR above our intended value of 15 at 3800$\mathring{A}$ showed that the mean age of the inner halo is 9.3 ± 1.4 Gyr using the Cummings et al. (2018) IFMR and MIST isochrones, or 10.8 ± 0.6 Gyr using the relation from Kalirai (2012).

Since the high-mass measurements could not be definitely attributed to a bias, I considered potential scenarios for their production. I found that ejected thin disk objects alone can not explain the sample, however, a combination of white dwarf mergers, halo blue stragglers, or increased star formation as a result of an interaction with the Sagittarius dwarf galaxy could explain this sample.

6.1.4 Upcoming Surveys

Ultimately, more work will need to be done to uncover a larger sample of halo white dwarfs. The conclusions of Chapter 4 led me to wonder whether future surveys will be able to definitely solve the issue of determining the early star formation of the Milky Way using white dwarfs.

Chapter 5 studies the future of white dwarfs as tracers of Milky Way evolution by simulating four major upcoming wide-field imaging surveys: The Legacy Survey of Space and Time (LSST), Euclid, the Roman Space Telescope's High Latitude Survey (HLS), and CASTOR. In order to do this, the Model presented in Chapter 2 was modified to produce synthetic white dwarf photometry in the bands and sky areas for each survey.

The results show that LSST, at its final 10-year depth, will detect more than 150 million WDs in its WFD survey, however, identifying them will be difficult at the faint end due to contamination from background galaxies and other stellar sources. Thus, future studies will have to rely on proper motions to identify the white dwarfs as, at equal colours, the faintness of white dwarfs means that they will lie closer to the Sun than other sources. This results in the white dwarfs experiencing a larger proper motion. By combining colours and proper motions, LSST will be able to select 7 million white dwarfs with greater than a 5σ proper motion measurement. Within this sample, a few hundred thousand will belong to the Galactic halo, improving the currently known sample by two orders of magnitude.

Figure 5.8 shows that the resulting sample from LSST will be able to detect the turn-off in the halo white dwarf luminosity function for the first time. This will allow for the most accurate age determination of the inner halo using white dwarfs, while simultaneously allowing for a determination of the star formation history of this population in a manner similar to the one presented in Chapter 3.

6.2 Future Work

Throughout this book, many persistent issues with our current understanding of white dwarf properties have been discussed. Future studies employing white dwarfs to study the formation and evolution of the Milky Way will need to tackle these problems in order to reduce the systematic uncertainties associated with current methods. This book will conclude with a brief discussion of a few nagging problems and the potential for future breakthroughs.

6.2.1 The Initial-to-Final Mass Relation

The initial-to-final mass relation (IFMR) is crucial in our understanding of stellar evolution as it determines the total amount of mass loss during post-main-sequence evolution and sets the initial conditions for the white dwarf. The IFMR is typically calculated using star clusters as the constant age of the stars means that the main-sequence turn-off mass can be related to the masses of their recently formed white dwarfs. An ideal dataset would include an even distribution of initial masses, however, observing old, low-mass, as well as very young, high-mass, clusters has proven to be challenging. One option at the low-mass end is halo globular clusters that typically lie at large distances from the Sun making the white dwarfs fainter than current instrumental sensitivities, although wide double degenerate binaries can also be used.

As Chapter 5 discussed, future observatories/surveys like the Mauna Kea Spectroscopic Explorer (MSE), the extremely large telescopes, and even the James Webb Space Telescope will be crucial to better sample the low-mass IFMR. Such programs will begin with imaging of the cluster to create a colour-magnitude diagram that can be used to identify the young, hot, white dwarfs. Follow up spectroscopy will need to target either the Balmer lines, like in Chapter 4, or the Paschen series in the infrared in order to determine the masses. Ideally, a handful of white dwarfs can be observed for each cluster to increase the statistics and to also study whether, for a given initial mass, the same final mass is achieved. Chapter 5 showed that this will be achievable in Messier 10 using LSST imaging combined with ELT spectroscopy, and there exists a number of other globular clusters at similar distances to M 10 (e.g, 47 Tucanae, Omega Centauri, M 12, M 71) that will serve as ideal candidates to expand this sample size.

6.2.2 White Dwarf Evolution and Binarity

Another open question in the study of white dwarfs that pertains to the studies presented in this book is the binary fraction. As mentioned in Chapter 4, the binary fraction of local white dwarfs is nearly half that of local main-sequence stars. A potential solution that has been proposed to solve this issue is a merger on either the main-sequence or during the double degenerate phase, but a disruption of a binary system when the primary star undergoes mass-loss along the post-main-sequence phase could also disrupt the system (El-Badry & Rix, 2018). Holberg et al.(2016) also suggests that the difficulty in detecting systems like Sirius, where the white dwarf has a massive primary companion, could also lower the discrepancy in the binary fraction.

The results of binary interactions will be evident in the mass distribution of white dwarfs. Previous observations have found three peaks in the white dwarf mass distribution (e.g, Kleinman et al., 2013) at <0.5 M$_\odot$, at ~0.6 M$_\odot$, and at ~0.8 M$_\odot$. The first peak, at <0.5 M$_\odot$, has been attributed to mass transfer in binary systems. The second peak is expected given single white dwarf evolution models, and is evident in the outputs of the model presented in Chapters 2 and 3. Finally, the third peak has had multiple explanations. First, these could be the results of merger events during or after the main-sequence that would produce more massive white dwarfs than expected. This scenario has the advantage of also explaining the discrepancy in the binary fraction as previously discussed. Another scenario, however, is that the IFMR flattens at intermediate mass due to a second dredge-up along the post-main sequence that decreases mass-loss Marigo et al. (2020). Finally, these white dwarfs may have experienced a delay in their cooling as a result of crystallization and the gravitational settling of heavy elements (Tremblay et al., 2019; Cheng et al., 2019).

Ultimately more data will be needed to study these scenarios and this will involve more high signal-to-noise spectroscopy combined with broadband photometry to properly characterize and potential companion objects. This will be possible with future surveys like MSE and WEAVE combined with space-based surveys like Euclid, the Roman Space Telescope, and CASTOR.

6.3   Final Thoughts

With the impending arrival of the next generation of deep, wide-field, astronomical surveys, the data needed to improve upon the studies presented in this **book** will be available. The ability to increase the sample of known white dwarfs by multiple orders of magnitudes will inevitably lead to a number of future studies involving both improvements to previous measurements as well as new studies that were previously unattainable. As detailed in Chapter 5, the turn-off in the halo white dwarf luminosity function will be observed in the LSST 5σ proper motion sample. The turn-off will allow for the age of the population to be determined for the first time, while the shape of the luminosity function can be used to determine the star formation history in a similar manner to the procedure presented in Chapter 3. The ability to use white dwarfs as tracers for Galactic evolution will move beyond just the local solar neighbourhood and into some of the more distant globular clusters in the Milky Way's halo and bringing with it a variety of new and exciting science cases.

www.ingramcontent.com/pod-product-compliance
Lightning Source LLC
LaVergne TN
LVHW042148190726
843493LV00006B/1567